MANUAL OF
SPECTROFLUOROMETRIC
AND
SPECTROPHOTOMETRIC
DERIVATIVE EXPERIMENTS

MANUAL OF
SPECTROFLUOROMETRIC
AND
SPECTROPHOTOMETRIC
DERIVATIVE EXPERIMENTS

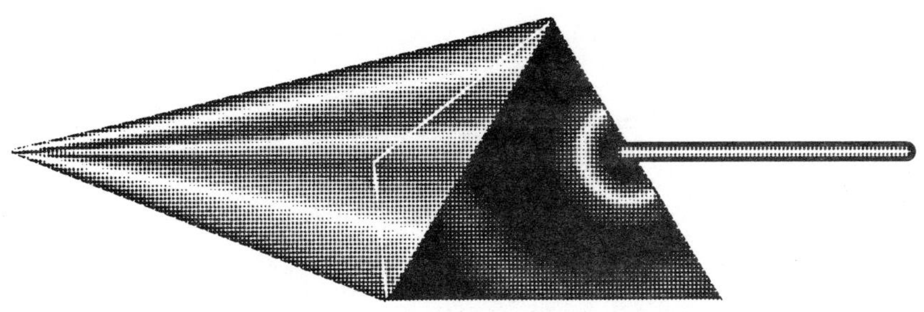

ALLESIA M. GILLESPIE, JR.

CRC Press

Boca Raton   Ann Arbor   London   Tokyo

Library of Congress Cataloging-in-Publication Data

Gillespie, Allesia M., 1928-
  Manual of spectrofluorometric and spectrophotometric derivative
experiments / by Allesia M. Gillespie, Jr.
     p.   cm.
  Includes bibliographical references and index.
  ISBN 0-8493-6390-X
  1. Spectrophotometry--Laboratory manuals.   2. Fluorescence
QP519.9.S58G55   1993
615'.1901--dc20   93-26029
                                                                       CIP

This book contains information obtained from authentic and highly regarded sources. Reprinted material is quoted with permission, and sources are indicated. A wide variety of references are listed. Reasonable efforts have been made to publish reliable data and information, but the author and the publisher cannot assume responsibility for the validity of all materials or for the consequences of their use.

Neither this book nor any part may be reproduced or transmitted in any form or by any means, electronic or mechanical, including photocopying, microfilming, and recording, or by any information storage or retrieval system, without prior permission in writing from the publisher.

CRC Press, Inc.'s consent does not extend to copying for general distribution, for promotion, for creating new works, or for resale. Specific permission must be obtained in writing from CRC Press for such copying.

Direct all inquiries to CRC Press, Inc., 2000 Corporate Blvd., N.W., Boca Raton, Florida 33431.

PREFACE

While doing the experiments found in this manual, it is suggested that the following books on luminescence and electronics be read:

- *Practical Fluorescence Theory, Methods and Technique,* by George G. Guilbault, Marcel Dekker, New York, 1973.

- *Absorption of Light and Ultraviolet Radiation Fluorescence and Phosphorescence Emission,* by George H. Schenk, Wayne State University, Allyn & Bacon, Boston, 1973.

- *Fluorescence and Phosphorescence Analysis Principles and Applications,* Edited by David M. Hercules, Interscience Publishers, New York, 1966.

- *Fluorescence Analysis: A Practical Approach,* by Charles E. White and Robert J. Argauie, Marcel Dekker, New York, 1970.

- *Digital Electronics For Scientists,* by H. V. Malmstadt and C. G. Enke, W. A. Benjamin, New York, 1969.

- *The Bugbook 1, Logic & Memory Experiments Using TTl Integrated Circuits,* by Peter R. Rony, David G. Larson, and Robert A. Braden, E & L Instruments, Inc., Derby, CT, 1974.

- *Basic Electronics,* by Van Valkenburgh, Nooger & Neville, Inc., John F. Rider Publisher, Inc., New York, 1969.

- *Basic Electronics For Scientists,* 2nd ed., by James J. Brophy, McGraw-Hill, New York, 1970.

This manual contains a wide assortment of luminescence and spectroscopic experiments that involve zeroth, first, or second derivatives. Emphasis is placed on those experiments that reveal important relationships and basic concepts.

The first few experiments are used to review basic concepts in electronics for the student. It is not necessary for the student to have had formal training in electronics to do these experiments as long as the books on electronics are utilized. Each experiment is characterized by these features:

1. A brief Introduction section that supplies background information and sets the stage for the experiment

2. Concisely stated Objectives that help the student focus on his/her reasons for being in the laboratory

3. Explicit Procedures that help guide the student efficiently and help achieve his/her objectives

4. Clear illustrations of the apparatus and physical set up

5. A list of the Instrumentation, Solutions, and Reagents required for the experiment that can be prepared and set out in advance of the laboratory period

6. Well-organized data tables listing in logical sequence the data to be recorded and calculations to be made

7. A Discussion section designed to help the student interpret the results of the experiment and to develop further or to illustrate an application of a concept

8. A Question section containing directions and questions designed to help the student evaluate his/her understanding of the experimental work by using the concepts and applying the principles associated with the experiment to the new situations

These laboratory experiments will enhance any course in electronics, fluorescence, or spectroscopy by giving the student practical experience that supports the theoretical lectures.

ACKNOWLEDGMENTS

I would like to thank the people who helped in the completion of this work, principally Dr. James Howell of Western Michigan University, who was the science advisor at the Food and Drug Administration during the time this research was done. He gave me invaluable help in setting up the instruments for these experiments.

I would also like to thank my late father and mother, my wife (Nell), my children (Otis, Allesia III, and Cynthia), and all my sisters (Doris Wright, Dr. Anna Marie Hayes, Gloria Green, Johnny Ethel Nichols, Delores Patricia Humphrey, and Deborah Gillespie).

TABLE OF CONTENTS

Section IV
Modern Derivative Spectrofluorometers and
Spectrophotometers 159

Section I

Application of Derivative Techniques to Drug Analysis

EXPERIMENT 1
OPERATIONAL AMPLIFIERS

Introduction

The name *operational amplifier* is derived from the theory of feedback amplifiers. The performance of the circuit of direct current (DC)-coupled amplifier with very high gain and with very heavy negative feedback around the amplifier is dependent on the causes of the feedback and the feed-in. If the operational amplifier has a great deal of gain, the gain is determined by the resistor ratio. For example, with a 100-k input resistor, a 10 gain amplifier is produced.

A ramp generator or integrator is produced by using a capacitor for feedback because the capacitor has to charge and discharge in response to input currents. There are a number of operations an integrator can perform, including waveform generation, mathematically finding the area under a curve of time waveform, and acting as a low filter.

If a great deal of resistors and capacitors are used with the feedback and feed-in units, an integrator is produced that can act as a high-performance active filter high pass, a band-pass, a band reject, or an equalizer.

Figure 1 shows that the operational amplifier consists of three gain units: a differential input unit, a high gain intermediate unit, and a high-power, low-impedance output unit. The differential input unit is the most important because it accepts input signals at very high impedance and with light loading, and accepts the difference between two input signals. The high gain intermediate unit provides high amounts of gain. The output unit provides drive power.

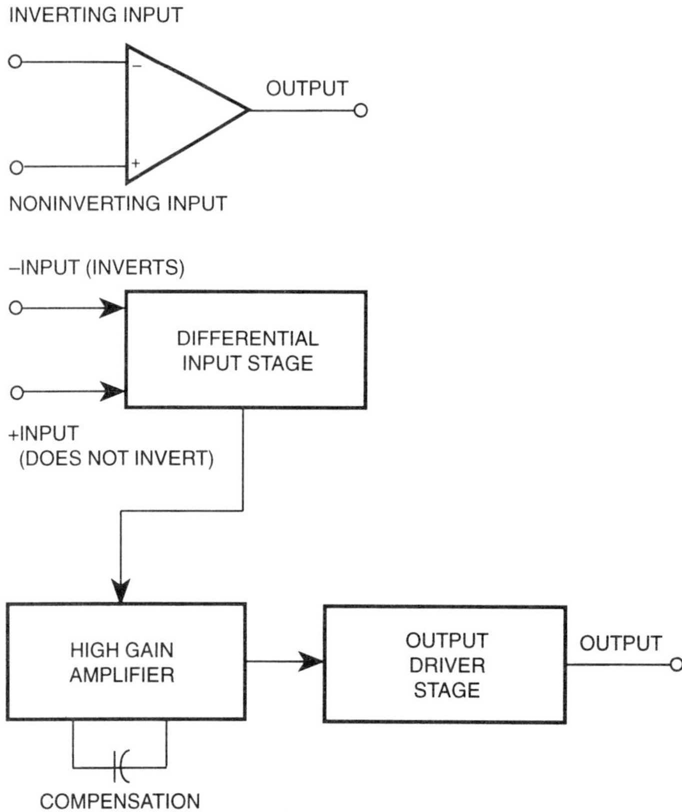

FIGURE 1   Components of an operational amplifier.

The differential input stage has two inputs, the inverting input $(-)$ and the noninverting input $(+)$. Positive supply lines are indicated by $V+$ and negative supply lines by $V-$. The operational amplifier has no ground connection. The circuit works with split positive and negative supplies, i.e., $+20$ and -20. A grounded signal reference lies between the positive and negative voltages.

A positive output is given when a small positive voltage is applied to the noninverting output $(+)$. If a positive voltage is applied to the inverting input (-), a negative output results. The circuit of the operational amplifier enhances the differences between the two inputs.

To be in a differential mode, both inverting and noninverting inputs are used. When one of the inputs is grounded (a reference that is halfway between the two supply voltages), a single-ended mode results. This experiment is conducted primarily to ensure that the operational amplifier is being used correctly.

The operational amplifier, when used as an integrator, has very high gain. It can possess gains above 10,000 when used without an external feedback. Lack of external feedback is also called open loop.

The operational amplifier most often is used with negative feedback. This negative feedback comes from having an external feedback resistor (RF) from the output back to the inverting input terminal (INV). This is called closed loop (Vcl). The amount of gain in this loop can be varied by altering the strength of the RFs.

This experiment asks the student to see if the operational amplifier has sufficient current to drive a loudspeaker. There are two categories of integrated circuits: (1) the digital integrated circuit, and (2) the linear integrated circuit, of which the operational amplifier is an example. The operational amplifier can replace circuits that are composed of many transistors, and is relatively inexpensive and reliable.

Objectives

The objectives of Experiment 1 are to ensure that the operational amplifier device used is in good operating condition, to become familiar with the breadboard and the ways in which to wire it and to become familiar with resistors, capacitors, etc.

Instrumentation, Solutions, and Reagents

Acquire an operational designer breadboard (E & L Instruments, Inc., 61 First St., Derby, CT 06418). Obtain the following from any electronic supply store: a 0.005-µF capacitor or equivalent; two 270 and 680 kΩ resistors; an 8-pin dual in-place (DIP) operational amplifier (#741C); an 8 Ω loudspeaker with switch; and a 9-V power supply.

Procedure

With the power off, connect the circuit of Experiment 1 (see the wiring diagram shown in Figures 2 through 4). Connect the outside row of the breadboard to ground (GND). Adjust the power supply to ±9 V by connecting V − on the diagram to −9 V and V + on the diagram to +9 V. If the E & L Instruments' breadboard is used, this adjustment can be accomplished by altering the dual variable supply. Insert the 741C DIP operational amplifier as shown in the diagram. The match is the reference point on the DIP. Complete the circuit using the appropriate wires, resistors, and capacitors (see Figures 2 through 4). Energize the loudspeaker

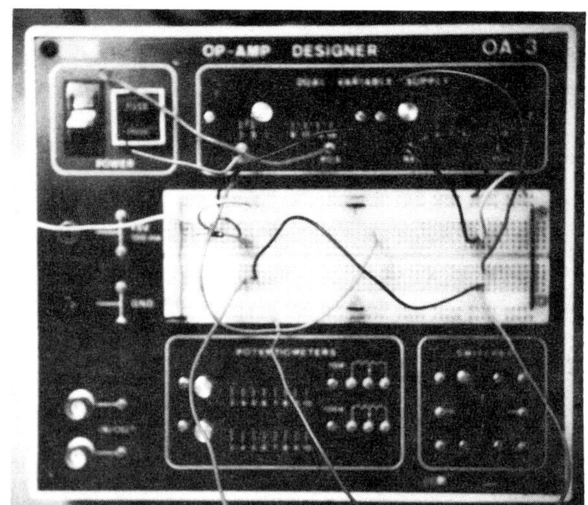

FIGURE 2   The operational amplifier breadboard.

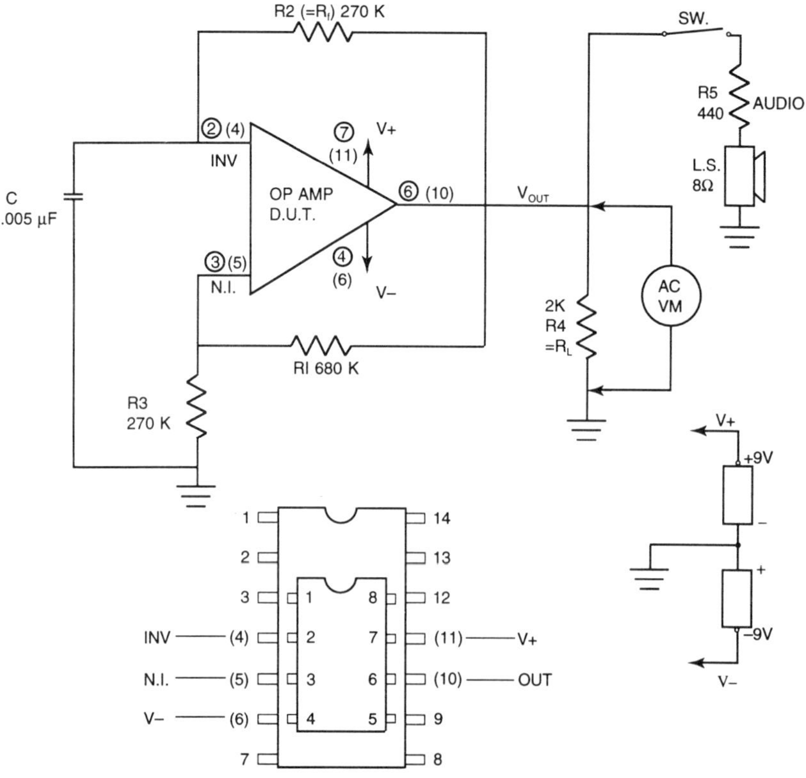

FIGURE 3   Schematic circuit of quick check test for operational amplifiers; pin arrangement (inset) that accommodates 8-pin packages.[19]

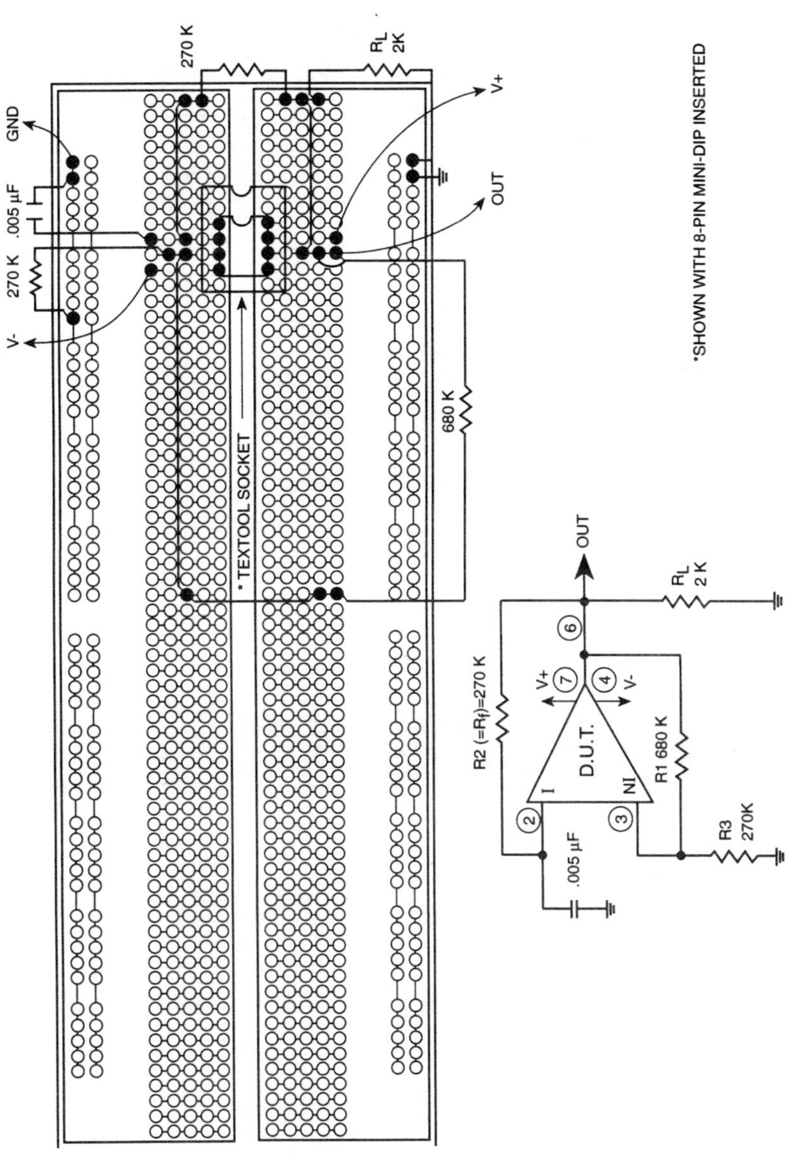

FIGURE 4 Wiring diagram for operational amplifier test.

by closing the switch to produce an audible output. The production of the audible tone checks the ability of the operational amplifier to supply reasonable output current.

Discussion

The operational amplifier device is operated on a stable multivibrator (oscillator) circuit. Using the loudspeaker (in series with 330 Ω) as a load draws a significant amount of output current with a corresponding lower output voltage. The production of an audible tone is a convenient check of the working condition of the operational amplifier.

Questions

1. What are capacitors?

2. What is the purpose of a capacitor?

3. What is a breadboard?

4. Describe a voltmeter and its purpose.

5. List the differences between a dual in-line and a TO case operational amplifier.

6. What are resistors and how is the ohm value of a resistor determined?

7. In relation to the operational amplifier, what is ground?

Answers to these questions may be found beginning on page 175.

EXPERIMENT 2
DESIGNING AN OPERATIONAL AMPLIFIER CIRCUIT THAT WILL GIVE ZEROTH, FIRST, AND SECOND DERIVATIVES ON A SPECTROFLUOROMETER

Introduction

The Model 194 operational amplifier manifold is suitable for performing experiments, breadboarding, and teaching with operational amplifiers. The Model 194 includes a dual ± 15 V, 100 mA power supply, an on-off switch, fuse, line cord, and mounting hardware. It will fit five amplifiers or less of the popular seven-pin ''Q'' case configuration in sizes up to a maximum $1\frac{1}{2}$ in. square. Adapters can be used for integrated circuit amplifiers. A balance resistor or potentiometer can be used with each amplifier.

The panel design is organized to give maximum interconnection flexibility. The input and output of each amplifier are arranged schematically on the panel. All adjacent connection points are equilateral triangles on $\frac{3}{4}$-in. spacing, and input and feedback component are mounted on dual banana plugs. The banana plugs can be plugged into the schematic as needed and the performance of the design may be demonstrated.

The theory that explains derivative luminescence is very similar to that which describes derivative spectroscopy. Derivative spectrometry was utilized by Griese and French in 1955[2] to obtain the separation of two spectral lines that emitted at the same wavelength. Researchers Collier and Singleton acquired the first and second derivatives of an intensity signal (obtained electronically)[3] as a function of time while the wavelength was changed linearly with time. The signal emitted nearly corresponded to the first or second derivative of intensity with respect to wavelength. This signal is a measurement of slope or curvature which causes the focus to be on the overlapping spectral lines and allows them to be separated. This application was used for other cases and reported in the literature.[4–6]

Derivative spectroscopy is applicable in those cases in which knowledge can be obtained through analysis of a spectral distribution. Any spectral distribution can be interpreted much more accurately at a given wavelength if the derivative is known along with the intensity of the spectral distribution at that wavelength. As increasingly higher derivatives are obtained, the more freedom from error is obtained when defining the spectral distribution. One example is a gas sample subjected to radiation of a particular wavelength in which the gas absorbs some of the radiation. First or second derivatives would give some qualitative information on the identity of the gas. Figure 5 illustrates the absorption of light by two different gases. Gas B absorbs radiation over a wide wavelength range;

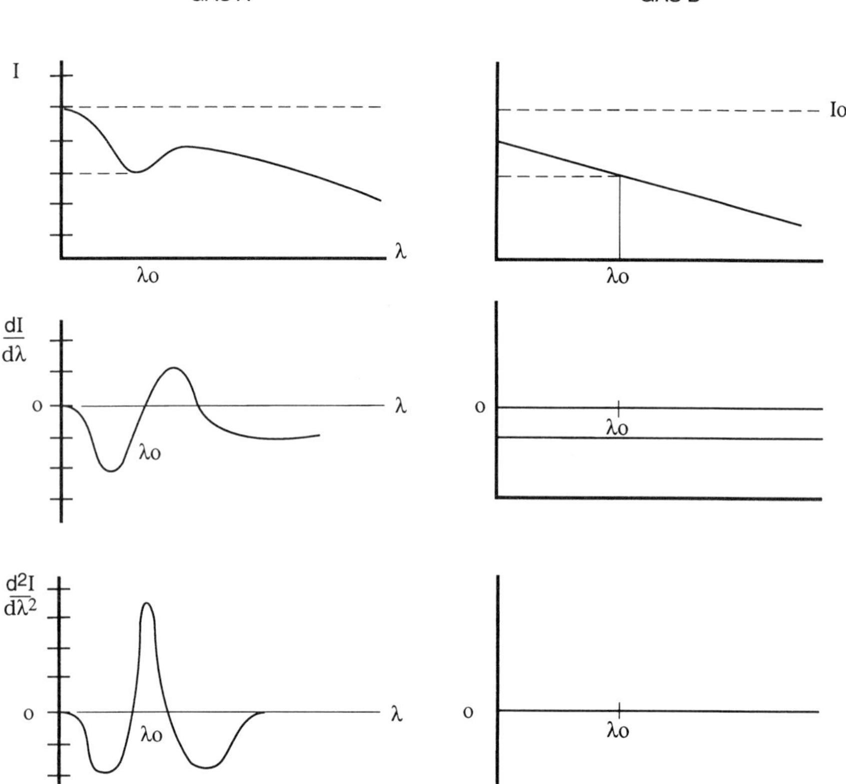

FIGURE 5   Zeroth, first, and second derivatives of an intensity signal as a function of time along with the change of wavelength with time.

therefore, the first and second derivatives of gas B are noticeably different from those of gas A, which absorbs radiation over a narrow bandwidth.

The theory of derivative spectroscopy embraces Beer's law:

$$A = \ln I_o/I = abc$$

where A is absorbance and a is the absorption coefficient that will change with wavelength. Assuming that a gas sample is being analyzed, c is the gas concentration. The derivatives of intensity are influenced by the absorptivity and intensity at a given wavelength. The first derivative is as follows:

$$dI/d\lambda - [dI_o/d\lambda \exp(-abc)] - [da/d\lambda - bcI_o \exp(-abc)] \tag{1}$$

$$dI/d\lambda/I = [IdI_o/I_o d\lambda - bcda/d\lambda] \tag{2}$$

The first part of this equation is a constant that indicates the sum total of slope in the radiation origin distribution. The second term changes linearly with the relative concentration of the gas.

The second derivative is as follows:

$$\frac{d^2I}{d\lambda}\Big/I = \left(\frac{I}{I_o} - \frac{d^2I_o}{d\lambda^2}\right) + \left(bc\,\frac{da}{d\lambda}\right)^2 - \left(\frac{2}{I_o}\frac{dI_o}{d\lambda}\frac{da}{d\lambda}\,bc\right) - \left(bc\,\frac{da^2}{d\lambda^2}\right) \quad (3)$$

The first unit of the equation is a constant measuring the curvature in the source of energy used to radiate the sample. If the bandwidth chosen to radiate the sample matches the maximum absorption wavelength, then the energy curvature will be at a maximum. At this wavelength, the slope will be zero, as shown in Figure 5, gas B. With a zero slope, the second and third term of Equation 3 becomes zero because these terms involve slope. With these two terms equal to zero, the fourth part of the equation becomes directly proportional to concentration.

Derivative spectrometers utilize an optical method called wavelength modulation to obtain derivatives. As the energy source transmits radiation to the sample, the wavelength is adjusted sinusoidally with time. The amplitude of the wavelengths used in the frequency modulation is about the same as the sum of the bandwidths in which the energy source irradiates.

In Figure 6, intensity is shown to change with wavelength. This produces a slope or the first derivative (see the slanting straight line in figure). The bottom wavelike line shows that radiation is being modulated (frequency changed) around the absorbing wavelength with respect to time. This causes modulation or change in the intensity of the radiation. This modulation is indicated by the top wavelike line and it has the same frequency as that of the wavelength modulation. The amplitude (height of the wavelength) of the intensity modulation that is manifested is directly related to the amount of change in intensity. When the radiation is taken through a detector, the intensity of the radiation will give a proportional electronic signal. The amplitude of the induced intensity modulation is therefore measured. This measured amplitude is directly related to the first derivatives of intensity with respect to wavelength.

Intensity again changes with wavelength in Figure 7, but instead of a sloped straight line, a curved line is produced (second derivative). The modulation of wavelength with time produces a modulation of intensity with time at twice the frequency (number of wavelengths per unit time). The more pronounced the curvature of the intensity distribution (second derivative), the more the amplitude of the induced intensity modulation increases. The mathematical theory that upholds these observations is beyond the scope of this manual.

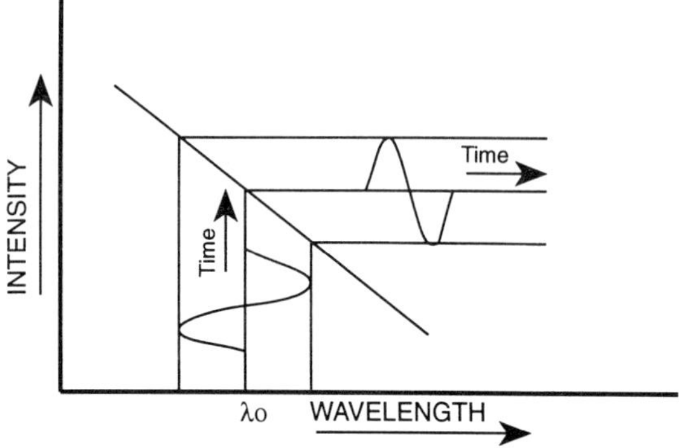

FIGURE 6   An intensity distribution which contains slope (first derivative).

Objectives

The objective in Experiment 2 is to design an operational amplifier circuit that will give zeroth, first, and second derivatives of a luminescence output and to record these derivatives.

Instrumentation, Solutions, and Reagents

Acquire a Model 194 operational amplifier manifold (Analog Devices, Cambridge, MA). From any suitable electronic supply store, obtain the following: 10 banana plugs; a 1.5-KΩ resistor; three No. 106B Analog Devices operational amplifiers; a 600-DC-size capacitor; a 10^7 and a 10^6 Ω resistor; four bias devices; some No. 22 solid wire; and assorted clamps. Next, obtain a cardboard box that is about 16 in. L $\times$ 13 in. W $\times$ 9 in. H to hold the instrumentation. Finally, you will need a Perkin-Elmer MPF 43A Spectrofluorometer with a Perkin-Elmer Model 123 recorder (or equivalent). Prepare a 0.157-ppm anthracene in methyl alcohol solution as follows:

1. Weigh 15.7 mg of anthracene (Aldrich Co. or equivalent), undried, into a 100-ml volumetric flask and dilute to about 60 ml with spectro-quality-grade methyl alcohol.

   - Heat in a steam bath for about 30 min until all the anthracene dissolves. Cool to room temperature and dilute to 100 ml with methyl alcohol.

2. Place a 1.0-ml aliquot (about 157 μg) into a 100-ml volumetric flask.

   - Evaporate to dryness under air.
   - Dilute to 100 ml with spectroquality methanol and shake well to give a 0.157 μg/ml (ppm) solution.

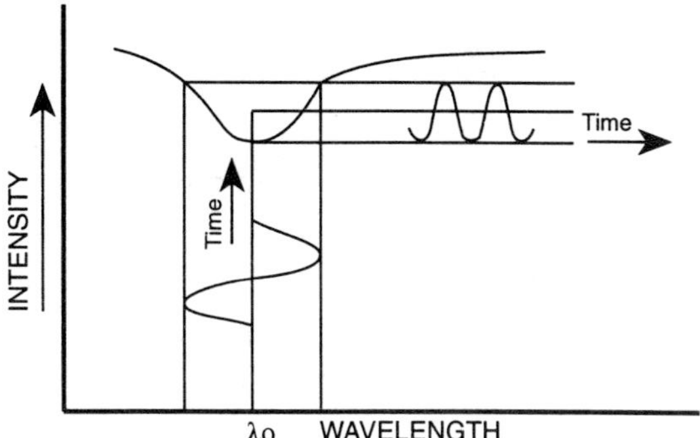

FIGURE 7　An intensity distribution containing curvature (second derivative).

Procedure

Connect the operational amplifier up to the Perkin-Elmer MPF 43A Spectrofluorometer so that its output is connected to the input of a differentiator circuit constructed from commercially available operational amplifiers. A circuit constructed from Analog Devices Model 194 operational amplifier manifold and Model 106B operational amplifiers is shown in Figure 8. Obtain the zeroth, first, and second derivatives of methyl alcohol and of 0.157 ppm anthracene in methyl alcohol.

Discussion

The modern Model 194 operational amplifier manifold is excellent for experimenting, breadboarding, and teaching. This model includes a dual ±15 V, 100 mA power supply, on-off switch, fuse, line cord, and mounting hardware. It accepts up to five amplifiers of the seven-pin "Q" case configuration in sizes up to 1½ in. square. Adapters are available for integrated circuit amplifiers. Provisions are made for a balance resistor or potentiometer for each amplifier.

The panel design is thoroughly organized to provide the highest degree of interconnection flexibility. The inputs and outputs of each amplifier are put in proper order schematically on the panel. All adjacent connection points are equilateral triangles on 3/4-in. spacing, and are input and feedback components of dual banana plugs. These are plugged into the schematic as desired, and the performance of the design can be demonstrated.

The zeroth, first, and second derivative spectra can be seen in Figures 9, 10, and 11 for methanol solutions of anthracene.

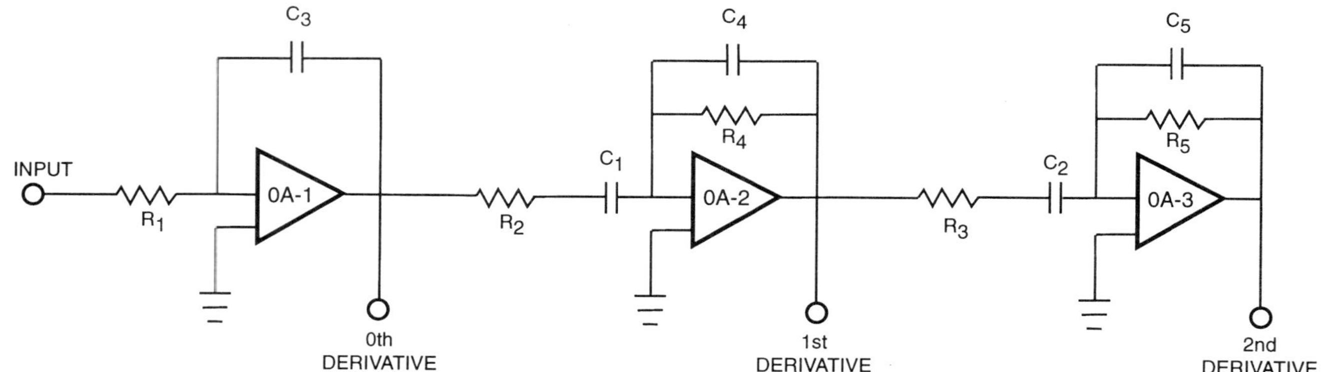

FIGURE 8   Circuit diagram of electronic differentiator 106B (Analog Devices, Cambridge, MA). OA-1 (left), OA-2 (center), OA-3 (right). R_1 = 1.5 kΩ, R_2, R_3 = 1 MΩ, R_4 = 10 MΩ, R_5 = 4.7 MΩ, C_1, C_2, C_3 = 1.5 μF, C_4 = 0.1 μF, and C_5 = 0.22 μF.

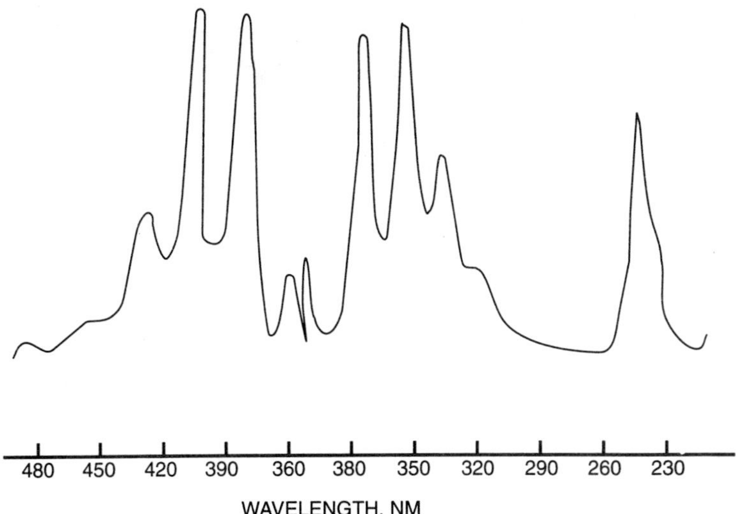

FIGURE 9   Normal excitation and emission spectra of 0.157-ppm anthracene. Emission scan: λ_{ex} = 357.5. Excitation scan: λ_{em} = 400.

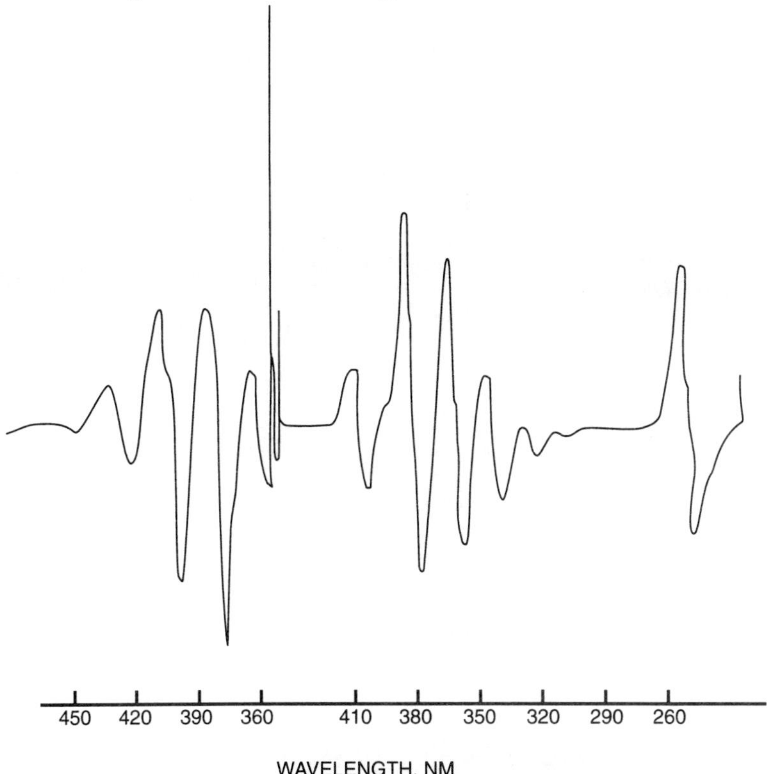

FIGURE 10   First derivative excitation and emission spectra of 0.157-ppm anthracene. Emission scan: λ_{ex} = 357.5. Excitation scan: λ_{em} = 399.

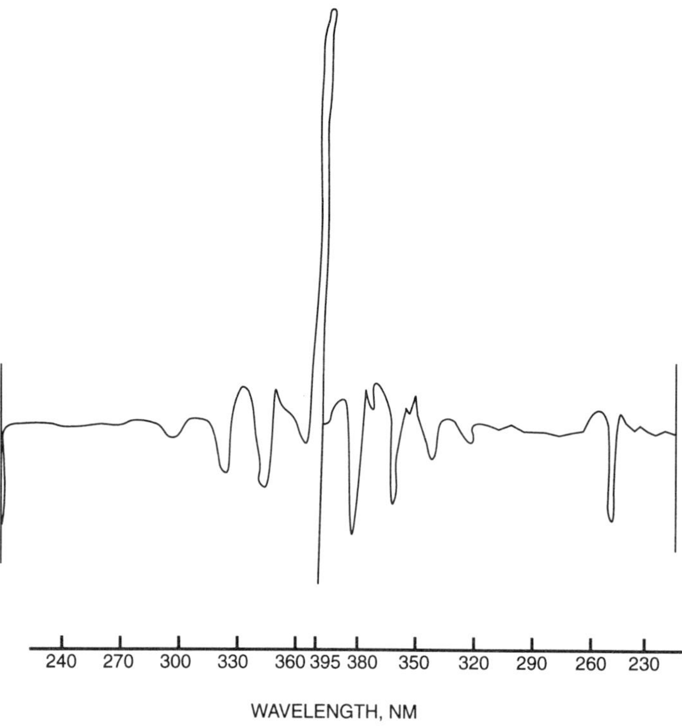

FIGURE 11   Second derivative excitation and emission spectra of 0.157-ppm anthracene. Emission scan: $\lambda_{ex} = 357.5$. Excitation scan: $\lambda_{em} = 399$.

Questions

1. *What are the first and second luminescence derivatives?*

2. *What is the purpose of an amplifier?*

3. *How does an amplifier work?*

4. *How does an operational amplifier work?*

5. Draw a block diagram of an operational amplifier.

6. Define the term impedance.

7. What is the difference between the inverting and the noninverting input?

8. What is "ground" for the operational amplifier?

9. How do we operate in a differential mode with the operational amplifier?

10. How do voltage amplifiers work? What are the essential principles that cause an input to be increased at the output? (Use the vacuum tube as an example.)

11. What is a grid?

12. What is a plate?

Answers to these questions may be found beginning on page 176.

EXPERIMENT 3
DETERMINATION OF THE ZEROTH, FIRST, AND SECOND DERIVATIVES OF THE LUMINESCENCE SPECTRA OF RESCINNAMINE AND RESERPINE

Introduction

Drugs that alter the human mind and behavior have received special attention since the beginning of recorded history. For example, alcoholic beverages and opiates have been utilized for this intention since ancient times. The discovery of lysergic acid diethylamide (LSD) in 1943 greatly awakened interest in psychotomimetic drugs. Mescaline and LSD appear to have similar properties.[7] Chlorpromazine has been used in the management of mental ailments. Several other psychotomimetics were developed a short time after the discovery of LSD, including the monoethylamide of lysergic acid, lysergic acid diethylamide with a bromine substituted in the 2 position, and lysergic acid hydroxybutylamide.

Psychopharmacology is the science concerned with drugs used to alter the mental state, and, thus the deportment of animals, including humans. A psychopharmacologic agent is a drug that repeatedly causes these modifications when used in what is judged to be a low-risk dosage. In addition, true psychopharmacologic agents will alter the state of mind and the manner in which an individual conducts him/herself in a foretellable way. When used for a serious mental malfunction, a satisfactory psychopharmacologic drug does not contribute to dependence and the patient taking the drug still makes an effort to adapt to their environment in a way that attains the desired ends.

Hypotensive or antihypertensive drugs are primarily used for the alleviation of high blood pressure. The most numerous types of high blood pressure are classified as essential (a milder form of hypertension) and malignant (a more severe form). The reasons for high blood pressure are not completely known at this time. There are a few cases in which ailments of the brain or spinal cord elevate blood pressure.[8] Kidney infections, toxemic pregnancies, and coarctation of the aorta have been physical causes. In such situations, when the physical problem has been mitigated, the blood pressure returns to a normal value. In nearly every case, doctors cannot explain the cause for the high blood pressure. This is referred to as *essential hypertension.* Essential hypertension is for the most part quite slow on onset. *Malignant hypertension,* on the other hand, has a rapid onset.

In cases of essential hypertension the patient is told to alter their diet, to exercise, etc., but is ordinarily not medicated. Malignant hypertension, however, is usually treated by drugs, most often by the rauwolfia alkaloid.

This is a reliable drug used in very serious cases.[9] The drugs used in ganglionic blocking may be united with the rauwolfia alkaloids. Rauwolfia lacks most, but not all, of the side effects exhibited by the ganglionic blocking agents. Chlorthiazide or similar diuretics may be included in the systematic course of medication and increase the actions of the antihypertensive drugs.

Objectives

The objectives of Experiment 3 are to determine the zeroth, first, and second derivatives of the luminescence spectra, and to compare the first and second derivative spectra with consideration for the suitability for quantitative analysis.

Instrumentation, Solutions, and Reagents

This experiment can be conducted using any of the modern spectrofluorometers that possess derivative function; e.g., the Hitachi Instruments, Inc. Model F-4010 is a fully computerized fluorescence spectrophotometer with derivative function capability. Obtain a spectrofluorometer that can produce zeroth, first, and second derivatives.

Prepare rescinnamine, 5 ppm, in methanol by:

1. Weighing 10 mg rescinnamine and placing it into a 100-ml volumetric flask. Dilute to 100 ml with absolute methanol to obtain a concentration of about 1 mg/10 ml.

2. Take 10 ml (1 mg) into a 200-ml volumetric flask and dilute it to 200 ml with methanol. This gives a 5-ppm (microgram/milliliter) solution.

Prepare reserpine, 5 ppm, in methanol by using the following method:

1. Weigh 10 mg reserpine into a 100-ml volumetric flask. Dilute to 100 ml with absolute methyl alcohol to obtain a concentration of approximately 1 mg/ml.

2. Put a 10-ml (1 mg) aliquot into a 200-ml volumetric flask and dilute to 200 ml with methyl alcohol. This gives a 5-ppm solution.

Procedure

Follow the directions given in the manufacturer's manual and obtain the zeroth, first, and second derivative of 5 ppm rescinnamine and 5 ppm reserpine in methyl alcohol using a λ excitation (λ_{ex}) of 312.5 nm and a λ emission (λ_{em}) of 402.5 nm for rescinnamine, and a λ_{ex} of 299.5 nm and λ_{em} of 350 nm for reserpine.

Prepare a table using the following headings:

Zeroth, First, and Second Derivative Bands in Methanol

Compound	Derivative mode	λ_{ex} (nm)	Emission scan λ_{em} peaks (nm)		Relative intensity	λ_{em} (nm)	Excitation scan λ_{ex} peaks (nm)		Relative intensity
			Major	Minor			Major	Minor	

Discussion

In this examination of a hypothesis, the zeroth, first, and second derivatives of reserpine and rescinnamine have been acquired. Reserpine is obtained from various *Rauwolfia* sp. The genus *Rauwolfia,* natural order Apocyanacae, comprises nearly 50 species that grow in tropical and semitropical geographical areas of South America, Africa, and Asia. The genus name recognizes a German physician and botanist of the 16th century, Leonard Rauwolf, who performed a detailed examination of medicinal plants in Africa[10] and Asia. His early writings briefly discuss the employment of *Rauwolfia* as a treatment for snakebites and scorpion stings. The drug was believed to be capable of use in the treatment of lunacy and as a sleep inducer in children.

Pharmacologic and chemical examinations of *Rauwolfia* were undertaken by the Indian chemists Bose and Sen, who managed successful hospital outpatient trials of the drug in 1941. Siddiqui and Siddiqui separated the first crystalline alkaloid from the plant in 1931. About 21 compounds have been isolated from *R. serpentina*. These compounds are printed in a table in Reference 10. Rauwolfia medicines come in the form of powdered whole root extracts, blends of chosen alkaloids, and the crystalline alkaloids reserpine and rescinnamine, free from adulterants. The structures of reserpine and rescinnamine are:

Reserpine

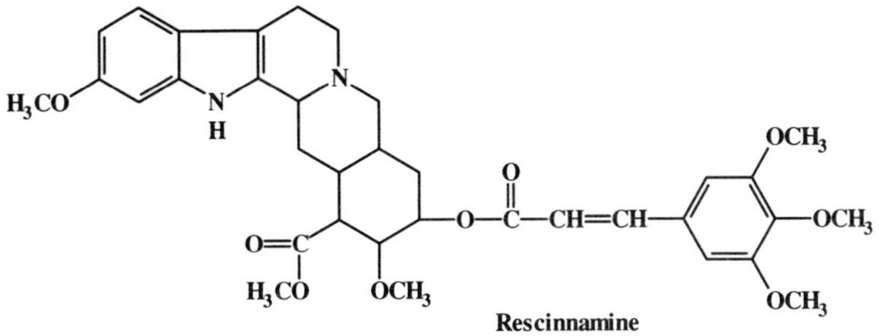

H_3CO ... OCH_3 ... $O{=}C{-}CH{=}CH{-}$ ... OCH_3 ... OCH_3 ... OCH_3 ... $O{=}C$ ... H_3CO ... OCH_3

Rescinnamine

The most obvious operations of the rauwolfia alkaloids are in the cardiovascular and central nervous systems. They are also utilized as antihypertensive agents and for psychotherapy.

Reserpine has several functional groups on its rings, but the most common is the methoxide group, which, in general, has very little effect upon wavelength or intensity of emission unless steric factors are involved. In a number of aromatic compounds, the substitution of methyl or ethyl groups caused substantial increases in the phosphorescence to fluorescence intensity.[11]

Rescinnamine not only has the methoxide group ($-OCH_3$) attached to the ring, but also differs from reserpine in that the attachment of the 1,2,3-methoxybenzene to the multiple ring system occurs via a $-OC{-}CH{=}CH{-}$ linkage in rescinnamine, whereas the linkage in reserpine is $-O{-}C{-}$.

The effect of more than one substituent on fluorescence is thus a resultant effect upon the mobility or freedom of the π electrons.[12] Weak *meta*-directing groups such as $-SO_3H$ and $-SO_2NH_2$, in combination with strong *ortho-para*-directing groups such as $-NH_2$ or $OH-$, can often increase fluorescence by increasing the freedom (one could say chaos) of the π electrons. A strong *meta*-directing group such as $-NO_2$ will tend to diminish fluorescence. Benzoic acid, which contains the *meta*-directing $-COOH$ group, is nonfluorescent or weakly fluorescent, but in combination with $-OH$ or $-NH_2$ groups, fluorescence can be increased.

For carbonyl compounds, the n-π* transitions occur at lower energy than π-π* transitions, which in turn occur at lower energy than σ-π* transitions.[13] For compounds not containing nonbonding electrons, the π-π* transition occurs at the lowest energy when compared to the other two transitions. The energies of the highest filled orbital relative to the lowest unfilled available orbital are important concepts for complete understanding of luminescence.

The n-π* and π-π* excited states are the major types found in organic chemistry. The n-π* excited state occurs in molecules having unbound electron pairs and generally exhibit weak fluorescence due to rapid intersystem crossing; they may also exhibit phosphorescence. The n-π^* transition is less intense than the π-π^* transition and, hence, n-π^* excited singlet states have longer lifetimes than π-π^* excited singlet states. This longer lifetime promotes intersystem crossing. A smaller energy difference also exists between the n-π^* singlet and triplet states, which also induces intersystem crossing and accounts for the phosphorescence generally associated with this excited state.

Both reserpine and rescinnamine contain −N− or −NH− in the ring. When −NH−, or −S− occur in the heterocyclic system, there is a tendency for them to contribute to the π electron system. Such rings are called π *excessive* and the order of contribution is −NH > −O− > −S−; therefore, fluorescence is increased.[12]

In polycyclic aromatic systems such as those found in reserpine and rescinnamine, the number of π electrons available is greater than in benzene and, thus these compounds and their derivatives are usually much more fluorescent than benzene.

Questions

1. *From a comparison of the zeroth, first, and second derivative excitation spectra of reserpine and rescinnamine, what derivative and wavelength are suitable for an analysis of an equal mixture of the two drugs?*

2. *From inspecting the spectra, can derivative luminescence be used for qualitative identification? How?*

3. *If only a filter fluorometer is available, name the method used to pick the correct excitation and emission filter for the second derivative maximum emission peak of reserpine.*

4. *Name at least two metabolites of reserpine.*

5. *Name several analytical problems that could be solved by the derivative luminescence technique.*

6. *What is an oscilloscope?*

7. *What are the uses for an oscilloscope?*

8. *Define the term* phase angle *or* phase difference.

9. *List two different methods of determining phase angles by instruments.*

10. *In what way can a cathode ray tube (CRT) be useful in a television set?*

Answers to these questions may be found beginning on page 179.

EXPERIMENT 4
DETERMINATION OF THE EFFECTS OF SOLVENT ON THE ZEROTH DERIVATIVE OF LUMINESCENCE

Introduction

Solvent polarity may play a significant role in influencing luminescence transitions. The interaction of solvent dipoles with n electrons tends to lower their energy, sometimes to the point that π-π^* transitions become the lowest energy transitions.[14] Because excited molecules are generally more polar than the same molecules in the ground state, interaction with a polar solvent may be more significant in the excited state than in the ground state.

Hydrogen bonding affects the wavelength of fluorescence emission because it tends to cause the molecule to absorb a lower amount of energy in order to move to a higher excited state and, therefore, the wavelength of emission is increased. This is due to the following relationship:

$$E = 10^7 \text{ nm} - \text{cm}^{-1}/\lambda \tag{4}$$

where E is energy and λ is the wavelength.

Aromatic and protonated amines, phenols, and thiols tend to be more acidic in the excited state than in the ground state.[16] Aromatic carboxylic acids, the carbonyl oxygens in aromatic aldehydes and ketones, and aromatic ring nitrogens tend to be more basic in the excited state than in the ground state.[16]

The luminescence of carboxyl, hydroxyl, and amino-substituted aromatics and nitrogen heterocyclics is particularly sensitive to the acidity of the solution in which they are dissolved.[17] Consideration of the structures of such compounds indicates that cations, neutral species, and anions may be present in the solution under various conditions. Some or all of the species may luminesce and the luminescence may be at the same or different wavelengths for the various species. The analyst must know the pH value at which maximum fluorescence is acquired for the cations, anions, and neutral compounds examined, and the analyst must be able to control the pH as a means of determining the type of compound present. A plot of fluorescence intensity vs. pH, which is called a titration curve, supplies the answer to this problem. The molecule in the excited state may display very different chemical behavior than the same molecule in the ground state.

The excited singlet state has a lifetime of 10^{-8} to 10^{-9} s, but some reactions, including protonation and dissociation, may be fast enough to compete with fluorescence emission. The 2-naphthyl ammonium cation is a stronger acid in the excited state than in the ground state. A higher K_a (acid dissociation constant) indicates a stronger acid. The estimation of pK_a^* values are based on the Forster cycle:

$$pK - pK^* = hc(\Delta Ea - \Delta E_f)/2.303KT$$

where pK = Negative logarithm of K_a
*   = Excited state
h = Planck's constant
c = Speed of light
ΔEa = Difference in energy of the O–O absorption band
ΔE_f = Difference in energy of the O–O fluorescence band

$$K \quad = \text{Boltzmann's constant}$$
$$T \quad = \text{Absolute temperature}$$

The intrinsic character of the absorbing and emitting anion, cation, or neutral species for a multitude of organic compounds depends on pH. If the excited state pK* differs from the ground state pK, excited state reactions may occur, in which case the structure of the emitting compound may not be the same as the absorbing compound. This is not quite accurate when describing phosphorescence because phosphorescence is seen in a rigid environment at low temperature. Under these circumstances the mobility of the saturated proton is reduced and prototropic reactions may not be able to compete with phosphorescence emission.

Objectives

The first objective of Experiment 3 is to show how fluorescence may vary with solvent and how this may be related to factors such as the dielectric constant of the solvent, the association of solvent and solute because of hydrogen bonding, the quenching of fluorescence by solvent molecules, and the ionization effects. The second objective is to use reserpine and rescinnamine as index compounds and to plot the wavelength of maximum fluorescence emission as well as the fluorescence intensity vs. the dielectric constant of several solvents.

Instrumentation, Solutions, and Reagents

Obtain one of the spectrofluorometers that is capable of obtaining the zeroth, first, and second derivative spectra (see Section IV, "Modern Derivative Spectrofluorometers and Spectrophotometers").

Obtain the following solvents:

Solvent	Dielectric solvent[18]
Methanol	32.6
Ethanol	24.3
1-Propanal	20.7
1-Pentanol	13.9

Prepare rescinnamine, 5 ppm, in each of the solvents as follows:

1. Weigh 10 mg rescinnamine into a 100-ml volumetric flask and dilute to 10 ml with the desired solvent to obtain a concentration of about 1 mg/10 ml. The exact concentration should be known.
2. Take 10 ml (1 mg) into a 200-ml volumetric flask and dilute to 200 ml with the desired solvent to obtain a 5-ppm (microgram/milliliter) solution.

Prepare reserpine, 5 ppm, in each of the solvents in the following way:

1. Weigh 10 mg reserpine into a 100-ml volumetric flask and dilute to 100 ml with the desired solvent to obtain a concentration of about 1 mg/10 ml.
2. Take a 10-ml aliquot (1 mg) into a 200-ml volumetric flask and dilute to 200 ml with the desired solvent to obtain a 5-ppm solution.

Procedure

Follow the manufacturer's instructions in order to operate the spectrofluorometer. Use the prepared 5-ppm reserpine and rescinnamine for samples. Start off with the methanol solutions in which the wavelength of excitation and the wavelength of emission of reserpine is 299 and 350 nm, respectively, while rescinnamine has a λ_{ex} of 312.5 nm and a λ_{em} of 427 nm. Obtain the zeroth, first, and second derivatives of reserpine and rescinnamine in methanol, ethanol, 1-butanol, and 1-pentanol.

Prepare a table using the following headings:

Solvent Effects on Zeroth Derivative Emission Spectra

Compound	Solvent	λ_{ex} (nm)	Emission scan λ_{em} peaks (nm) Major	Minor	Relative fluorescence intensity
	Methanol				
	Ethanol (95%)				
	1-Butanol				
	1-Pentanol				

Discussion

A solvent may act to dissolve in more than one way. When an ionic salt is dissolved in water, the process of solution involves separation of the cations and anions of the salt with accompanying positioning of solvent molecules about the ions. Such positioning of solvent molecules about the ions of the solute is possible only when the solvent is extremely polar and the dipoles of the solvent are drawn to and held by the ions of the solute. The arrangement of solvent molecules about solute ions is called *solvation* for solvents other than water, but when water is the solvent, the term used is *hydration*. The solvent must also possess the properties necessary to keep the solvated charged ions apart with the least amount of energy. The dielectric constant gives an idea of how well a solvent is able to keep charged ions apart. The higher the dielectric constant, the easier it is for the solvent to separate electrostatic charges in the solvent. In this manner, water, which has a dielectric constant of 80, reduces the force of attraction between two opposite electrical charges in water to $^1/_{80}$ of the force between the two charges in a vacuum. The dielectric constant of a vacuum is taken as 1. The dielectric constants of some common liquids are given below:[19]

Hydrogen cyanide	116
Water	80
Glycerin	46
Ethylene glycol	41
Methyl alcohol	33
Ethyl alcohol	25
n-Propyl alcohol	22
Acetone	21
Aniline	7.0
Chloroform	5.0
Ethyl ether	4.3
Benzene	2.2
Carbon tetrachloride	2.2
Octane	1.9

Liquid hydrogen cyanide and water possess the greatest ability to dissolve electrolytes. Electrolytes are substances whose aqueous solutions will conduct an electrical current because the ions of the electrolyte carry the current. Arrhenius acids, bases, and salts are examples of electrolytes. However, octane possesses the least ability to dissolve an electrolyte and, thus, has a dielectric constant of only 1.9. In most cases, the large dielectric constant for water is due to hydrogen bonding. Nonpolar solvents such as octane, carbon tetrachloride, and benzene are not capable of dissolving electrolytes.

A polar liquid such as water may exhibit solvent action through its ability to break a covalent bond and to cause ionization of the same solute. For instance, hydrogen fluoride dissolves in water and acts as an acid as a consequence of the following reaction:

$$HF + H_2O \rightarrow H_3O^+ + F^-$$

The ions formed when the covalent bonds of the solute are broken are maintained as ions in the solution by the same mechanism of hydration that acts on ionic salts such as NaCl.

Another method by which a polar liquid may act as a solvent is the situation in which the solvent and solute are able to be joined by means of a hydrogen bond arrangement. The solubility of the low molecular weight alcohol in aqueous solution is ascribed to the sufficiency of power of the alcohol molecules to become part of a water-alcohol complex:

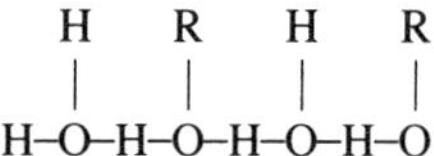

As the molecular weight of the alcohol becomes larger, it becomes less polar and less able to rival the water molecules for a position in the combination of atoms formed by hydrogen bonding. For this reason, high molecular weight alcohols are inadequately soluble in aqueous solutions. When the number of carbon atoms in a straight chain alcohol is five or more, solubility in water is greatly diminished. However, branching in the carbon chain of the alcohol tends to augment the solubility in water. Also, if an alcohol has more than one hydroxyl group, a greater degree of water solubility exists. The reason that high molecular weight compounds such as sugars, gums, and glycosides are soluble in water is the presence of numerous hydroxyl groups on the compounds.

The hydrogen bond also is mainly responsible for the solubility of ethers, aldehydes, ketones, acids, and anhydrides in water or other polar solvents such as methanol. The complex that forms between these compounds and water is similar to that illustrated with alcohol and water. The hydrogen bonding of the ethers, aldehydes, and ketones occurs through the hydrogen atom of water and the oxygen atom of these compounds. For examples, the hydrogen bonding of acetone occurs as follows:

$$(CH_3)_2CO + HOH \rightarrow (CH_3)_2CO-H-O-H$$

Just as five carbon atoms appear to be the number of carbon atoms that must be present before alcohol solubility is diminished in water, the same appears to be true of ethers, aldehydes, and ketones.

The hydrogen bond forms between the hydrogen atom of one compound and the oxygen, fluorine, or nitrogen atom of another compound. This is because oxygen, fluorine, and nitrogen have the highest electronegativities of the elements and, therefore, possess a vigorous electrostatic attraction toward the hydrogen ion. When the hydrogen bond strengths are ranked, fluorine, the most electronegative element, forms the strongest hydrogen bond. The strength of the hydrogen bond is directly related to the electronegativity of the atoms forming the bond, with highly electronegative atoms forming stronger bonds. Thus, the order of hydrogen bond strengths is as follows:

$$H-F < H-O < H-N$$

Nitrogen forms the least strong bond with hydrogen. For this reason, compounds such as the amines (RHN_2) are less soluble in water than alcohols (R–OH).

Nonpolar liquids generally cannot dissolve polar compounds because they are unable to form the dipoles needed to overcome the force that strives to draw ions of an ionic salt close together, and they cannot break a covalent bond to form ions or to form strong hydrogen bonds. In general, like dissolves like and nonpolar solvents will only dissolve nonpolar compounds. The weak force of the bonds found in nonpolar substances is generally van der Waal's force. An example of a nonpolar solvent dissolving a nonpolar compound is an oil (e.g., soybean oil) dissolving in hexane. In a few cases, a polar compound will dissolve in a nonpolar solvent. For example, it is found that isopropyl alcohol will dissolve in hexane. It is theorized that the polar isopropyl alcohol, which has permanent dipoles, induces a dipole in the nonpolar hexane, and thereby temporarily causes it to become a polar solvent. Here, the negative end of the isopropyl alcohol is attracted to the positive end of the hexane and vice versa, and solvation is accomplished. This particular type of force is called a *permanent dipole-induced dipole force*. Induced dipoles occur when the positive nuclei of one atom is attracted to the negative electrons of another atom. The electrons are in constant motion, and at some point in time, the positive nuclei of one atom are close enough to the electrons of another atom to be attracted to each other.

As previously stated, like dissolves like. The more similar of the atomic structures of the solute and solvent, the more soluble the solute is in the solvent. For example, water is highly soluble in methanol (HOH and CH_3OH), but water is insoluble in oil. Oil, of course, has a vastly different structure than water.

Organic substances that hold polar aggregations that are able to bond with the hydrogen or oxygen atoms of water are soluble in water as long as the molecular weight of the chemical substance is relatively small. $-OH$, $-CHO$, $-COH$, $-CHOH$, $-CH_2OH$, $-COOH$, $-NO_2$, $-CO$, $-NH_2$, and $-SO_3H$ are polar functional groups. When these groups are attached to an organic molecule, the solubility of the organic compound in water is enlarged. Hydrocarbon radicals or other nonpolar or weakly polar functional groups tend to lessen the solubility of the organic compound in water. The solubility of organic compounds in water is also affected by other factors such as the number of carbon atoms in the hydrocarbon radical, decreased solubility being associated with an increased number of carbon atoms and the presence of halogens. Halogens decrease solubility in water because they are nonpolar and because they increase the molecular weight. The more numerous the polar functional groups present in a particular compound, the more soluble in water the compound. The position of a polar functional group also appears to affect solubility. For example, *m*-dihydroxybenzene has greater water solubility than *o*-dihydroxybenzene, which, in turn, is more soluble than *p*-dihydroxybenzene.

High molecular weight compounds are usually insoluble or only slightly soluble in water. Substances with high melting points have a strong intermolecular force of attraction, and this force acts to reduce solubility in water.

Geometric isomers can be of the *cis*- or *trans*-type. The *cis*-type is more soluble in water than is the *trans*.

Questions

1. *Why don't ionic compounds dissolve to an appreciable extent in such solvents as benzene or carbon tetrachloride?*

2. *Taking into account the results of this experiment, what was the effect on the wavelength of excitation and emission of reserpine as the dielectric constant of the solvent increased?*

3. *Based on the experiment, what was the effect on the wavelength of excitation and emission of rescinnamine as the dielectric constant of the solvent increased?*

4. *Examine the structure of reserpine and rescinnamine and discuss possible reasons why the greatest relative fluorescent intensity for reserpine occurred in 1-butanol, while the greatest relative intensity for rescinnamine occurred in methanol.*

Answers to these questions may be found beginning on page 181.

EXPERIMENT 5
WAVELENGTH AND INTENSITY OF SECOND DERIVATIVE EXCITATION SPECTRA AS A FUNCTION OF SCAN SPEED

Introduction

In order to fluoresce, a compound must first absorb sufficient energy so that it can move to a higher energy state. Therefore, a discussion of some of the aspects of derivative spectroscopy is necessary. UV-visible absorption spectra are characterized by the broad nature of the observed bands, due to the extension over and above numerous rotational, vibrational, and electronic transitions. The resolving capacity of commercial instruments is often 100 times less than the natural bandwidth of the compounds being investigated.

The resolution of any spectroscopic instrument is defined as $\lambda/\Delta\lambda$, where $\Delta\lambda$ is the point or line of division of the wavelengths of two lines barely recognized as two, and λ is the average wavelength in nanometers.

In the actual performance or application, the resolution is fixed by many causative agents.[20] These agents include:

1. The standard used to determine whether two lines are separated.
2. The nature of the light. The light could be on-axis and may fully illuminate the dispersing appliance. In addition, the light could be coherent (a point source at a distance without lenses), incoherent (as though the slit was self-lighting), or a mixture of coherent and incoherent.
3. The nature of the optical imaging constituent parts, e.g., the types of lenses and mirrors used.
4. The nature of the dispersing device.
5. The aberrations of the total optical outline.
6. Aligning the optical instrument for optimal focus.
7. The slit width used.
8. The resolution of the photographic plate detector, the resolution of the viewing lens, the viewing eyepiece of a spectroscope, the exit slit, and photodetector of a spectrometer.
9. The strength of the radiation of the two lines being resolved.
10. The natural widths of the spectral lines emitted by the energy source of the spectroscopic instrument.

The absorption of ions and molecules is ascertained by setting the solution in a glass or silica cell and then putting the cell into a UV or visible spectrum spectrophotometer and quantifying the amount of radiant energy absorbed at each wavelength. An absorption band rather than a series of lines is ordinarily seen. Absorption bands can vary from 50 nm up to and exceeding 300 nm. An absorption band is made up of many absorption lines that are so close together that they cannot be detected as lines by the spectrophotometer. This is a sign that the excited state of a substance in solution is cleaved into several sublevels, n, with each sublevel having a small difference in its energy content. Each sublevel, therefore, is furnished with substances that absorb photons of different energies. For example, the absorption of light by a solution of $Cu(H_2O)_6^{+2}$ occurs over the wavelength range 495 to 700 nm; i.e., in the excited state ions will be found having energies ranging from those that absorb low energy photons found at 750 nm to ions that absorb high energy photons found at 495 nm.[21] Because the absorption bands are usually wide, most mixtures have overlapping absorption bands, and this overlap interferes with the quantitative analysis.

By determining the rate of change of absorbance with wavelength (the first derivative spectrum), the absorption bandwidth can be put into a manageable, reduced state. The fastest moving rates of change are found near the peak maximum, therefore, this relatively small part of the absorption band exhibits the most vivid changes in the derivative spectrum. Inherent in electronic instrumentation that produce first derivative spectra is a basic problem with respect to noise. The high-frequency noise produced by any spectrophotometer is a fast-altering signal. Therefore, an electronic derivative circuit will behave by enlarging the altering signal in relation to the smaller frequency signal due to the varying absorbance of the sample. For this reason, the majority of uncomplicated derivative devices that add to the effectiveness of the total electronic derivative instrument have less utility because of the highly augmented noise displayed on the resultant spectra.

One process by which derivative spectra are produced abstains from this problem, however, by using a dual-wavelength spectrophotometer. Devices similar to the Perkin-Elmer Model 556 optically create first derivative spectra. This prevents the addition of noise to the spectra. Perkin-Elmer has manufactured an electronic unit capable of producing both first and second derivative absorption spectra. Second derivative spectra turn broad peaks found in the absorption spectra into fine second derivative peaks. The second derivative spectra often are able to resolve overlapping absorption bands.

Additional operating parameters are required in second derivative spectroscopy, as compared to absorption or first derivative spectroscopy. Unlike the others, the second derivative accessory differentiates the varying absorbance signal vs. time, not wavelength. If the spectrophotometer possesses a fixed scanning speed then d^2A/dt^2 is equivalent to $d^2A/d\lambda^2$. The selection of scanning speed immediately influences the intensity of the second derivative spectrum. A faster scanning speed produces a more rapidly changing absorption signal to the derivative circuit as well as a greater result. On the other hand, to obtain an individual response, knowledge is lost at elevated speeds and the resolution of the spectrum is impaired. The most useful derivative spectra are recorded at comparatively swift speeds such as 60 nm/min or better.

Objectives

The objectives of Experiment 5 are to obtain the second derivative excitation spectra of standard solutions of reserpine and rescinnamine in methanol at scan speeds of 15, 30, 60, and 120 nm/min, and to determine the wavelength and fluorescent intensity of second derivative excitation maxima and minima at each scan speed.

Instrumentation, Solutions, and Reagents

Obtain a spectrofluorometer capable of obtaining zeroth, first, and second derivatives and that has various scanning speeds. Also, obtain rescinnamine, 5 ppm, and reserpine 5 ppm in methanol, in the manner described in Experiment 4.

Procedure

Follow the manufacturer's directions to prepare the spectrofluorometer for analysis and use the prepared 5 ppm of reserpine and 5 ppm of rescinnamine as samples.

When obtaining the zeroth derivative use the following λ_{ex} and λ_{em}: reserpine, λ_{ex} 299 nm and λ_{em} 353 nm; rescinnamine, λ_{ex} 312.5 nm and λ_{em} 427 nm.

When obtaining the first derivative use the following λ_{ex} and λ_{em}: reserpine, λ_{ex} 299 nm and λ_{em} 327 nm; rescinnamine, λ_{ex} 312.5 nm and λ_{em} at 402.5 nm.

When obtaining the second derivative use the following λ_{ex} and λ_{em}: reserpine, λ_{ex} 299 nm and λ_{em} 327 nm; rescinnamine, λ_{ex} 312.5 nm and λ_{em} 350 nm.

Obtain the zeroth, first, and second derivative excitation and emission spectra of methanol, reserpine, and rescinnamine at scan speeds of 15, 30, 60, and 120 nm/min.

Prepare a table using the following headings:

Zeroth Derivative Excitation Spectra

Compound	Solvent	Derivative peak monitored max λ_{ex}	Excitation peak height (chart units)	Max λ_{ex}	Spectrofluorometer instrument setting

Second Derivative Spectra

Compound	Sample sensitivity settings		Scan speed (nm/min)	Max. λ_{ex} of peak used	Net peak height (chart units)
	Coarse	Fine			

Prepare graphs plotting scan speed vs. relative excitation intensity (peak height).

Discussion

Increased scan speed tends to increase the slope of the zeroth order spectrum, which produces a more intense second derivative excitation peak. This is a direct consequence of time rather than wavelength derivatives being taken. Because wavelength is presumably a linear function of time, the time-wavelength correlation is supposedly linear.

Increasing the scan speed increases the sensitivity of the instrument so that, with no change in concentration, a particular change in slope will become a larger peak. This is analogous to increasing the sensitivity of a gas-liquid chromatography (GLC) instrument, usually accomplished by reducing the attenuation. However, an optimum point exists for increased sensitivity in derivative fluorescence as it does in GLC. For a given luminescence peak, information can be lost at higher speeds and a decrease in resolution occurs.

Questions

1. *Why does faster scanning speed result in a more intense fluorescent peak for a given slope?*

2. *How would increasing the scanning speed aid the analysis of a sample in which only a small amount of sample was available?*

3. *In the zeroth excitation and emission spectra, which peaks should give the greatest second derivative peaks?*

4. *If analysis was necessary for two fluorescent compounds having maximum peaks at the same wavelength, but different slopes in obtaining the peaks, how could the difference in slope along with increasing scan speed be used to find the compound with the greatest slope?*

Answers to these questions may be found beginning on page 182.

EXPERIMENT 6
INFLUENCE OF TURBIDITY ON FLUORESCENCE DERIVATIVE SPECTRA

Introduction

The accurate measurement of phenol in wastewater was one of the first useful examples in the relevant application of derivative spectroscopy to the qualitative and quantitative determination of a substance in the midst of a turbid solution. Ascertaining the amount of a substance via its first or second order derivative spectrum is justified by the Beer-Lambert law, which claims that turbidity does not influence linearity.[22]

Liquids used as thirst quenchers sometimes contain quinine. These quinine-containing liquids also often contain material extracted from fruits. This material causes the liquids to be turbid. The chemically extracted fruit substances absorb UV light, as does quinine. For this reason, direct spectrophotometric measurement of the quinine in this liquid is not possible. However, the measurement of quinine can be accomplished by using derivative spectroscopy, which is unaffected by the interfering UV-absorbing fruit materials as long as proper conditions are maintained.

The dark coloring used in the manufacture of cola drinks makes direct spectrophotometric determination of caffeine impractical. However, derivative spectroscopy has been used to produce spectra that are free from interfering absorbance.

Milk is turbid due to the presence of the fat emulsion. The determination of protein in milk is better accomplished with derivative spectroscopy than with the standard protein analysis method. The fat emulsion in milk exhibits the essential characteristics of a scattering sample in the area from 800 to 400 nm. In this region of the electromagnetic spectrum, the absorption spectra is due to the fine fat particles found in the milk. However, when a second derivative spectrum is obtained, the interfering background absorption is eliminated.[23]

Objectives

1. To use a 5-µg/ml stock standard solution of reserpine to make up standard solutions containing 0, 0.25, 0.5, and 1.0 g/l aluminum silicate in methanol, and to obtain the zeroth, first, and second derivative excitation spectra of each solution in order to determine the manner in which turbidity can influence spectra.

2. To repeat step 1 above, but substituting 2-µg/ml reserpine standards containing 0, 0.5, 1.0, 1.5, and 2.0 g/l cornstarch.

3. To make up reserpine standards in methanol containing 0, 2, 4, 6, and 8 ppm reserpine, then to take these standards and obtain second derivative excitation spectra and make up a standard curve. Also, we want to add aluminum silicate so that the concentration will be 2.0 g/l, and we want to repeat the standard curve as above in order to observe the effect of turbidity.

Instrumentation, Solution, and Reagents

Use a spectrofluorometer that can ascertain the zeroth, first, and second derivative spectra (see Section IV, "Modern Derivative Spectrofluorometers and Spectrophotometers").

Prepare four reserpine standards of about 5 µg/ml each containing aluminum silicate at 0, 0.25, 0.5, and 1.0 g/l as follows:

1. Weigh 10 mg reserpine into a 100-ml volumetric flask and dilute to 100 ml with absolute methanol to obtain a concentration of about 100 µg/ml. Pipette 5 ml (500 µg) each into four 100-ml volumetric flasks. Store the 100 µg/ml solution in the refrigerator.

2. Obtain aluminum silicate, reagent grade. Add the following amounts, respectively, to three of the 100-ml volumetric flasks containing the 500-µg reserpine solutions:

   - 25 mg; this gives a concentration of 0.025 g of aluminum silicate per liter of solution.
   - 50 mg; this gives a concentration of 0.050 g of aluminum silicate per liter per solution.
   - 100 mg; this gives a concentration of 0.1 g of aluminum silicate per liter of solution.

Prepare two other groups of standard solutions of reserpine for standard curve preparation by following these steps:

1. Pipette 0, 2, 4, 6, and 8 ml of the 100-µg/ml reserpine solution into separate 100-ml volumetric flasks and dilute to 100 ml with methanol to obtain 2-, 4-, 6-, and 8-ppm solutions.

2. For the second group of standard solutions, weigh out 200 mg each of aluminum silicate and place into five 100-ml volumetric flasks. Add 0, 2, 4, 6, and 8 ml of the 100-µg/ml reserpine standard solution and dilute to 100 ml with methanol. This will give solutions containing 0, 2, 4, 6, and 8 ppm reserpine and 2.0 g/l aluminum silicate.

Procedure

Follow the manufacturer's directions to set up and operate the spectro-fluorometer. Obtain the zeroth, first, and second excitation and emission derivative spectrum of reserpine, 5 ppm. Obtain the zeroth, first, and second excitation and emission derivative spectra of reserpine solutions (5 ppm) containing 0, 0.25, 0.5, 1.0, and 2.0 g/l aluminum silicate in methanol at a scanning speed of 60 nm/min. Rinse the cuvette well with methanol and obtain a solvent blank by obtaining the zeroth, first, and second derivative spectra of methanol using the same instrument conditions used for the reserpine solutions. Obtain the second derivative excitation of 0, 2, 6, and 8 ppm reserpine alone, and 0, 2, 6, and 8 ppm reserpine containing aluminum silicate at a concentration of 2 g/l in methanol using the following parameters: monitored λ_{ex} = 279 nm, λ_{em} = 353 nm, scan speed 60 nm/min, and scan range 245 to 345 nm. Run in duplicate and obtain a solvent blank by obtaining the second excitation derivative spectra of methanol under the same conditions and prepare standard curves.

Compare the results obtained to the zeroth, first, and second derivative spectra of 5 ppm reserpine alone in regards to the following: intensity of a dominant peak, wavelength of the first minima, wavelength of the maximum peak, wavelength of the second minima, and the angle of slope of the baseline at the beginning of the scan of the zeroth derivative.

Compare the second derivative excitation spectra obtained for the blank which contains 2 g/l aluminum silicate and to the second derivative excitation spectra of methyl alcohol which contain no aluminum silicate in regards to location and size of the peaks.

Use the results obtained to compare the second excitation derivative spectra of the solutions containing reserpine alone to the second excitation derivative spectra of the solutions containing reserpine and aluminum silicate with respect to the following: intensity, wavelength of the first minima, wavelength of the maximum peak, and wavelength of the second minima.

Examine the results. Is the following statement true: "the effects of increasing turbidity can be canceled almost completely and the resulting second derivative spectra are essentially identical"?

Prepare the following tables for the zeroth, first, and second derivatives of a methanol blank and a reserpine solution (approx. 5 ppm) containing 0, 0.2, 0.5, 1.0, and 2.0 g/l aluminum silicate. Always precede an aluminum silicate standard solution with a solution of reserpine of the same concentration but containing no aluminum silicate. This procedure allows precise comparison of the solutions.

First Derivative

Resperpine, conc. (ppm)	Aluminum silicate, conc. (g/l)	Net peak height (chart units) of max peak		λ (nm) of first minima		Comments
		Excitation	Emission	λ_{ex}	λ_{em}	
	0					
	0.25					
	0					
	0.5					
	0					
	1.0					
	0					
	2.0					

2nd Derivative

Resperpine, conc. (ppm)	Aluminum silicate, conc. (g/l)	Net peak height (chart units) of max peak		λ (nm) of first minima		Comments
		Excitation	Emission	λ_{ex}	λ_{em}	
	0					
	0.25					
	0					
	0.5					
	0					
	1.0					
	0					
	2.0					

Effect of Turbidity on the Maximum Wavelength and Intensity of Zeroth, First, and Second Luminescence Derivatives of Reserpine

Aluminum silicate conc. (g/l)	Derivative mode	Emission scan[a]	Intensity decrease (%)	Excitation scan[b]	Intensity decrease (%)
0	Zeroth				
0.25	Zeroth				
0.50	Zeroth				
1.0	Zeroth				
2.0	Zeroth				
0	First				
0.25	First				
0.50	First				

Effect of Turbidity on the Maximum Wavelength and Intensity of Zeroth, First, and Second Luminescence Derivatives of Reserpine (continued)

Aluminum silicate conc. (g/l)	Derivative mode	Emission scan[a]	Intensity decrease (%)	Excitation scan[b]	Intensity decrease (%)
1.0	First				
2.0	First				
0	Second				
0.25	Second				
0.50	Second				
1.0	Second				
2.0	Second				

Note: Peak intensity and wavelength relative to the standard containing no aluminum silicate. All scan speeds were 60 nm/min.

[a] Change in maximum emission wavelength in nanometers.
[b] Change in maximum excitation wavelength in nanometers.

Discussion

The effect of increasing background due to turbidity up to a certain limiting concentration can be partially canceled in the first and second derivative. The limiting concentration of turbidity, above which the first and second derivative cannot cancel out interference, must be determined individually on each turbid solution.

Light is reduced in intensity when it is passed through an intervening, interfering substance that contains scattered particles of different refractive indices. Because the amount of scattering is proportional to the number of scattering particles, the foundation of an analytical method is formed.[24]

The fraction of light conveyed in concentrated scattered particulate matter may be measured directly with a turbidimeter. Ordinarily, the light scattered at 90° is measured using a nephelometer or Tyndallmeter.

Turbidimetry and nephelometry are analytical procedures based on the dispersal of light by particulate matter in solution. In turbidimetry, the source radiation is conveyed directly through the sample solution and the decrease in intensity is measured. In nephelometry, the ray of light irradiating the sample is measured at a 90° angle to the point at which it is scattered by the sample particles.

Turbidimetry is best suited for determining relatively large concentrations of dispersed particles, whereas nephelometry is appropriate for the analysis of low concentrations.[25] The terms turbidimetry and nephelometry are, in general, confined to specific cases in which the concentration of the particulate matter in a sample is being determined. If the suspended particles are about the same size or smaller than the wavelength of the light striking them, the light is scattered, but if the particulate matter is larger than the wavelength of light hitting them, the light is reflected.

For nephelometric analysis, the suspended particles should be small in size in comparison to the wavelength of light used, so that scattering rather than reflection is of greater importance. The reason is that scattering is related to the size and shape of the pattern of the secondary rays in space, and this provides the best possible view of the intensity found at 90° to incident light. A lesser fraction of light is bent at right angles to the incident light due to nullifying interference with particles that are greater in magnitude. Particulate materials that are the same size or larger than the wavelength of the light used cause much of the dispersed or reflected light to move forward rather than at right angles or backward.

There is an optimum particle size for scattering. For wavelengths in the UV and visible regions of the electromagnetic spectrum the optimum particle size is between 0.1 and 1 μm (colloidal size).

The association between absorbance and concentration is unrelated to a line when a turbidimetric analysis is performed on large particulate materials because an increasingly larger fraction of the divergent light moves forward and still reaches the detector. The reason that this occurs is multiple reflections and forward angle scattering.

The quantity of light that diverges from the principal light source is contingent on the following: the quantity of the particles per unit volume, the ratio of refractive indices of the particulate matter and its encircling environment, the size and shape of the particulate material, and the wavelength of the light striking the sample solution. In order for reflection or scattering to occur, a quantitative difference must exist between the encircling environment refractive index and the particle refractive index.[26] It is often advantageous to change solvents for the purpose of increasing the difference between the solvent and the particles in the refractive index.

The most important quality effecting the achievement of turbidimetric or nephelometric analysis is the power to regulate the size and shape of the particulate matter. In turbidimetry, signaling does not rely solely on the weight of suspended material per unit volume, but on the light-blocking particles per unit volume and the cross-sectional area of the particles.

Suspended particles are almost never the same size. Thus, it is significant that all samples and standards have identical apportionments of small, medium, and large size particles. This means that samples and standards should be prepared under the same conditions as often as possible. In generating particulate matter via precipitation, the size of the particles may differ greatly according to the following factors: changes in the concentration of the reactants, temperature, shaking or mixing, pH, the presence of nonreactive substances, the sequence of the blending of the reactants, and the time given for growth of the particles. Deviations in particle growth are the most common error in turbidimetry and nephelometry, but would probably not cause problem in derivative fluorescence.

Surface active compounds are often combined with precipitating solutions to make the colloidal condition steadfast and to forestall the growth of large particles through agglomeration. The mechanism involved is absorption of the surface active compounds to the particle surface so as to slow down or stop particle growth and, thereby, to make particle size resistant to sudden change. The most frequently used surface active compounds are glycerol, gelatin, dextrin, and gum arabic.

Light wavelength is often more important in turbidimetric analysis than in nephelometric analysis because in turbidimetry one obligatorily prevents or minimizes absorption. This is not a significant factor in nephelometry. In measuring the light transmitted through a solution, no method is known of perceiving the difference between light energy that has been removed by absorption from light or that by scattering. For this reason it is greatly consequential in turbidimetry that a wavelength be chosen in an area in which the sample solution does not strongly absorb. If the sample solution is colored, strong absorbance can be obtained by using visible light of the same wavelength as the colored solution. For solutions that are free from substances that dim or obscure but contain dark particles, radiant energy at a wavelength corresponding to the red or near-IR spectrum may exhibit little absorption. In nephelometric analysis, where absorption is not really a source of trouble, UV light is more commonly used. In both turbidimetry and nephelometry, the standard solutions used to effect analysis should be the same color as the sample solution so as to reduce the effects of absorbance.

In turbidimetric analysis, the transmittance, T, is given by the equation $T = I/I_o$. I_o is the intensity of the incident light measured after passing through a blank cell containing the solvent being used, and I is the intensity of the radiation after passing through the solution being analyzed. The equation that associates the concentration, C, of dispersed particles to scattering, S, is

$$S = \log I_o/I = Kbc \tag{5}$$

where K is a proportionality constant or turbidity coefficient and b is the cell width. Another constant, the turbidity τ, equal to 2.303 K is often used. The value of the proportionality constant depends on the size and shape of the particulate matter as well as the wavelength and refractive index of the solution and the particles present. The turbidity equation is effective only for small particles for which Rayleigh scattering is the only mechanism for attenuation and for low concentrations of dispersions where multiple scattering is unlikely. Customary turbidimetric analysis involves preparing a standard curve and obtaining the scatter measurements for the sample and relating it to a given concentration.

In nephelometry, scattered intensity, I_s, is related to the concentration, C, of the dispersed particles by the equation:[28]

$$I_s = K_s I_o C \qquad (6)$$

where K_s is a constant and I_o is the incident light intensity. All determinations of quantities are made under similar conditions.

Conducting a quantitative nephelometric analysis requires the preparation of a standard curve whereby log I_s/I_o is plotted vs. the concentration of dispersed particles. Log I_s/I_o is then obtained for a sample solution and the concentration of the particles in the sample is determined.

Much of the instrumentation used in turbidimetry and nephelometry bears a close resemblance to the ordinary colorimetric or spectrophotometric devices found in most laboratories. The most common application of the light dispersing analysis is in determining the relative cleanliness of various water and beverage goods. However, light dispersing methods have also been used for the following analyses in liquid solutions:[29] sulfate as $BaSO_4$, carbonate as $BaCO_3$, chloride as $AgCl$, fluoride as CaF_2, cyanide as silver cyanide, calcium as the oxalate or oleate, and zinc as ferrocyanide. A widely used carbon dioxide determination involves bubbling the gas through an alkaline solution of a barium salt and then analyzing for the precipitated, particulate barium carbonate.

The analysis of foods and beverages include the analysis of sugar products containing sediment or foreign particles (syrups, molasses, etc.), citrus fruit juices, and turbid alcoholic beverages. The amount of benzene in alcohol can be determined by adding water to the solution as this causes the benzene to become immiscible. Alkaloids can be quantitated by forming a precipitated particulate suspension after the addition of phosphomolybdates. Turbidimetric analysis has been used to determine amino acids, vitamins, and antibiotics. Nephelometric analysis has been used for the quantitation of protein, yeast, glycogen, and of beta and gamma globulins in blood serum and plasma. A turbid solution of nicotine can be prepared by precipitation with potassium iodomercurate and this can be quantitated by light scattering methods.

Measurement of the amount of turbidity has been used to ascertain the endpoint in titrations. Some representative titrations include the titration of 10^{-2} *M* fluoride with calcium, 10^{-5} *M* bromide with silver, sulfate with barium, and ketones in alcohol-ketone mixtures with water.

Turbidimetry and nephelometry are used to automatically analyze, without interruption, air and water particulate matter that act as pollutants. This particulate matter could be dust or smoke in air as well as particle suspension in water.

Questions

1. For which type of turbid solutions would derivative spectrofluorometry be used?

2. Given that vol $= 4/3\pi(r)^3$, find the volume of a particle that has a radius of 0.1 μm.

3. Given that particles present in a solution have a volume of 6×10^{-15} cm^3 and density of 2 g/cm^3, calculate the number of particles found in 5 g of the particles in a solution.

4. What is the cross-sectional blocking area to a beam of monochromatic light passing through a solution containing 50×10^{13} particles in which each particle has a radius of 10^{-5} cm?

5. It is important to measure the concentration of a dissolved fluorescent solute, but present in the solution is a cloudy precipitate of some other solute. Filtration and centrifugation fail to remove this precipitate. Devise a means of determining the concentration of the dissolved fluorescent solute in the presence of the precipitate.

6. *Define Rayleigh and Raman scattering.*

Answers to these questions may be found beginning on page 183.

EXPERIMENT 7
ANALYSIS OF MIXTURES OF STANDARD RESERPINE AND RESCINNAMINE IN TABLETS USING DERIVATIVE LUMINESCENCE

Introduction

Second derivative fluorescence is intrinsically quantitative because the linear connection between sample emission and the concentration of the active ingredient is unaffected by differentiation. As long as the standard fluorescence exchanging rules of slit selection, scanning speed, etc., are acknowledged and a suitable mode has been singled out, quantitative measurements can be made by measuring from peak maximum to peak minimum. The final sensitivity is limited by noise.

One of the greatest benefits of derivative luminescence is in the analysis of mixtures, in which an interfering component produces a sloping background zeroth derivative luminescence upon the zeroth derivative fluorescent spectrum of the desired component. For example, chemists at Perkin-Elmer Corp. have used first and second derivative spectra to quantitate very small amounts of fluorescein and rhodamine B in 77 ppb eosin. The eosin peak at 540 nm (emission spectra) is the major peak, while the fluorescein and rhodamine B represent shoulders on the short and long wavelength sides of the eosin peak, respectively.[30] The first or second derivative scans can be used to quantitate these dyes.

Terharr and Poro[30] used both differential and derivative fluorometric techniques to determine anthracene in the presence of quinine sulfate. The quinine sulfate spectra were instrumentally removed, leaving only the differential spectra of anthracene. The anthracene spectra was composed of several overlapping bands. The first and second derivative spectra were obtained. A standard solution of 17 ppb anthracene in $N/10$ H_2SO_4 was made up, and the zeroth, first, and second derivative spectra were obtained and compared to the zeroth, first, and second derivative spectra of the differential spectra in the anthracene-quinine sulfate mixture. The result was that they were comparable both quantitatively and qualitatively.

The combined use of differential and derivative techniques increases the quantitative capability of fluorescence spectroscopy.

Objectives

The objectives of Experiment 7 are to create 1:1 mixtures of reserpine and rescinnamine tablets, and to quantitate both reserpine and rescinnamine using derivative fluorescence spectroscopy in order to demonstrate the utility of this technique as a means of analyzing mixtures.

Instrumentation, Solution, and Reagents

Obtain a spectrofluorometer with a zeroth, first, and second derivative capability (see Section IV). Prepare a 1:1 reserpine-rescinnamine standard solution in the following way:

1. Weigh 10 mg reserpine and 10 mg rescinnamine into a 10-ml volumetric flask and dilute to 100 ml with methanol to obtain a concentration of 100 μg/ml.
2. Pipette 0, 1, 2, 3, 4, 5, and 6 ml of the 1:1 reserpine-rescinnamine standard into a 100-ml volumetric flask and dilute to 100 ml with methanol in order to obtain 0-, 1-, 2-, 3-, 4-, 5-, and 6-ppm solutions.

Obtain authentic tablets (0.25 mg) containing reserpine alone and another set of tablets containing rescinnamine alone (0.25 mg per tablet). Prepare 5 ppm rescinnamine into a 100-ml volumetric flask and dilute to 100 ml with absolute methanol to obtain a concentration of 1 mg/10 ml. Take 10 ml (1 mg) into a 200-ml volumetric flask and dilute to 20 ml to obtain a 5-ppm solution. Prepare 5 ppm reserpine in methanol in the following way:

1. Weigh 10 mg reserpine into a 100-ml volumetric flask and dilute to 100 ml with absolute methanol to obtain a concentration of about 1 mg/10 ml.
2. Take a 10-ml aliquot of the solution prepared above and place it into a 200-ml volumetric flask and dilute to 200 ml with methanol to obtain a 5-ppm solution.

Procedure

Follow the manufacturer's directions, use only the scanning speed of 60 nm/min, and obtain the zeroth, first, and second derivatives of 5 ppm reserpine and rescinnamine in methanol.

Prepare a table using the following headings:

Zeroth, First, and Second Derivative Bands in Methanol

Compound	Derivative mode	λ_{ex} (nm)	Emission scan, λ_{em} peaks (nm)		Relative intensity λ_{em} (nm)	Excitation scan, λ_{ex} peaks (nm)	
			Major	Minor		Major	Minor
Reserpine	Zeroth	299	353				
	First	299	327				
	Second	299	327				
Rescinnamine	Zeroth	312.5	427				
	First	312.5	402.5				
	Second	312.5	350				

All scanning speeds should be set at 60 nm/min. First and second derivative peaks reported should be "negative" peaks. The baseline should be run at the 50% relative intensity chart unit of the recorder. Negative peaks are defined as those that deflect below the baseline, while positive peaks are those that deflect above the baseline. Use the following symbols to describe the spectra: "W" (weak) is less than 10 relative units; "M" (moderate) is between 10 and 30 units; "S" (strong) is greater than 30 units.

Analyze for reserpine by preparing a 1:1 mixture of reserpine and rescinnamine tablets so as to obtain a concentration of about 4 ppm each in methanol as follows:

1. Separately weigh 20 rescinnamine tablets and 20 reserpine tablets to obtain an average weight.

2. Separately grind the 20 reserpine and 20 rescinnamine tablets and pass the ground material through a #60 sieve.

3. Weigh out an amount of the reserpine and rescinnamine tablet material equivalent to about 400 μg each into a 100-ml volumetric flask and dilute to 100 ml with absolute methanol.

4. Weigh out an amount of the ground reserpine and rescinnamine tablets equivalent to about 200 μg of reserpine and 200 μg of rescinnamine into a 100-ml volumetric flask. Spike this sample by adding about 200 μg each of reserpine and rescinnamine standard in methanol. Dilute to 100 ml with absolute methanol. Shake all the solutions for about 30 min. Let the solutions set, occasionally shaking, for another 30 min. After this, filter the solutions through #40 filter paper, discarding the first 10 ml and holding the rest for analysis.

Prepare a standard curve by obtaining the second derivative excitation spectra of the 1:1 reserpine-rescinnamine standards at concentrations of 0, 1, 2, 3, 4, 5, and 6 ppm using the following parameters on the spectrofluorometer:

- Monitored wavelength (λ) = 276 nm
- Maximum λ_{em} = 353 nm
- Scanning speed = 60 nm/min
- Scanning range = 250 to 300 nm

Determine the net intensity by measuring the perpendicular distance from the second minima to the maximum peak and plot net intensity vs. concentration.

Using the same instrumental parameters, obtain the second derivative excitation spectra of the sample and the spike. Also use the standard curve or the 4-ppm 1:1 mixed standard to calculate the amount of reserpine in the sample.

Obtain the spectra of the 4-ppm 1:1 mixed standard both before and after each sample spectrum in order to detect any changes in instrument sensitivity during an analysis.

Analyze for rescinnamine in the following way:

1. Use the solutions prepared previously.

2. Prepare a standard curve by obtaining the second derivative excitation spectra of the 1:1 reserpine-rescinnamine standards at concentrations of 0, 1, 3, 4, 5, 6, and 8 ppm using the following parameters on the spectrofluorometer equipped with a differentiator circuit:

   - Monitored wavelength (λ) = 337 nm
   - Maximum λ_{em} = 426.9
   - Scan speed = 60 nm/min
   - Scan range = 290 to 360 nm

Determine the net intensity by measuring the perpendicular distance from the first minima to the maximum peak and plot net intensity vs. concentration.

Using the same instrument parameters, obtain the second derivative excitation spectra of the sample and the spike. Finally, using either the standard curve or the 4-ppm 1:1 mixed standard, calculate the amount of rescinnamine in the sample.

Run the 4-ppm standard both before and after each sample spectrum so as to detect any changes in instrument sensitivity during an analysis.

Discussion

Derivative techniques are useful for both quantitative and qualitative analyses of luminescent compounds that exist as mixtures. The choice of wavelengths and whether the first or second excitation or emission derivatives should be used in an analysis must be experimentally determined on an individual basis for the particular mixtures encountered.

Several factors must be considered when planning an analysis. The wavelengths and derivative chosen must be such that the excitation or emission peak used for analysis must be highly characteristic of only one compound. The first and second derivative excitation or emission peaks used for quantitation must not occur too close to one another because this interference may make it difficult to obtain a linear intensity vs. concentration plot. Ideally, a minimum of about 50 to 100 nm should separate the peaks for second derivative plots of well-resolved spectra. It is also essential to obtain linear intensity vs. concentration plots of the ingredients being analyzed.

The method has the advantage of sensitivity, speed, and simplicity. The major benefit is that quantitation can occur with no prior separation, even with the overlap of the excitation or emission spectra of the compounds, as long as appropriate instrument parameters and sample preparation steps are employed.

Questions

1. *Propose a derivative luminescent method for the analysis of a 1:1 phenanthrene-anthracene mixture in cyclohexane.*

2. *Name a natural product that contains phenanthrene.*

3. *Name some of the physical and chemical properties of phenanthrene.*

4. *Phenanthrene solutions exhibit a blue fluorescence. Using the information found in this experiment, to which wavelength does this correspond?*

5. *What is an important use of anthracene?*

Answers to these questions may be found beginning on page 185.

EXPERIMENT 8
THE USE OF DERIVATIVE SPECTROSCOPY IN THE IDENTIFICATION OF ORGANIC COMPOUNDS

Introduction

Ordinary spectroscopic methods have not been conducive to finding or analyzing overlapping absorption bands. Derivative spectroscopy has been of the utmost value in locating these absorption spectra and as an aid in their analysis. Using derivative spectrometry, it has been possible to make clearer changes in the curvature of absorption spectra. Many of these changes would largely go unnoticed if other methods were used. This procedure helps to clarify any minuscule differences between absorption spectra. Using derivative spectroscopy, a derivative spectra can be obtained in which maxima and minima points correspond approximately to the curvatures found in the absorption spectra. The baseline of the absorption spectra will also produce a maxima and minima on the derivative spectra. Wavelengths of the derivative and absorption spectra do not necessarily match up. The derivative spectra are functions of the change of amplitude with concentration and the scanning rate.

The Cary Model 118 spectrophotometer (Varian, Instrument Division, Cary Products, 611 Hansen Way, Palo Alto, CA 94303) is constructed for the purpose of measuring how much light of varying wavelengths is absorbed or transmitted by solid, liquid, or gaseous substances. By adding other devices to the Cary 118, several other kinds of measurements can be made including fluorescence, reflectance, and derivatives. Automation can be achieved by the addition of still other devices, and makes it possible to obtain readouts from a digital computer or printer, automatic repetitive scanning, and automatic sample changing. Light in the visible or near-UV region is directed to a mirror which, in a small fraction of time, will send the light through the sample and then rotate twice to send the light through the sample and then through a reference. The two light beams then intersect at a phototube. The light passing through the reference is approximately equal in intensity to the light striking the sample, and the relative intensities of the two beams are used to measure the magnitude of light that has been absorbed or transmitted. The reference contains the same solvent in which the sample is dissolved so that solvent absorbance is subtracted and only the sample absorbance is measured. In most cases, absorbance is proportional to concentration. When used to create derivatives, the rate of change of an absorbance spectra with respect to wavelength or time can be recorded.

Objectives

The objectives of Experiment 8 are to obtain derivative spectra of organic compounds in order to help identify them; and to become more familiar with the operation of a derivative spectrometer.

Instrumentation, Solutions, and Reagents

Obtain a Cary Model 118 spectrophotometer or equivalent with derivative attachment. (See Section IV, "Modern Derivative Spectrofluorometers and Spectrophotometers", for other spectrophotometers capable of obtaining derivative spectra.)

Prepare 9.5% HCl in ethanol using the following method:

1. Pipette 380 ml of concentrated HCl into a 4000-ml graduate.
2. Dilute to 4000 ml with 95% ethyl alcohol.

Obtain the following standard compounds: phenprocoumon, warfarin, and acenocoumaral. These compounds can be ordered from the U.S. Pharmacopeial Convention, Inc., 12601 Twinbrook Parkway, Rockville, MD 20852. Prepare phenprocoumon standard solution as follows:

1. Weigh 1 mg phenprocoumon into a 100-ml volumetric flask.
2. Dilute to 100 ml with 9.5% HCl in ethanol and mix well.

Prepare acenocoumaral standard solution using the following procedure:

1. Weigh 1 mg acenocoumaral standard into a 100-ml volumetric flask.
2. Dilute to 100 ml with 9.5% HCl in ethanol and mix well.

Procedure

Take each of the standard solutions of phenprocoumon, warfarin, and acenocoumarol and treat as follows using the Cary 118 spectrophotometer or equivalent (the following directions assume the use of the Cary 118):

1. Scan and record an absorption spectra (see the manufacturer's manual).
2. Return the scan to the original starting position, but use a fresh portion of the chart.
3. Set the "Function" switch to "Deriv." Lift up the pen so that it is held away from the paper.
4. Try different settings of the "ABS RANGE" switch "Scan Speed Selector" and run scans at the desired wavelengths. Select settings that give the optimum spectra.

5. Scan until the wavelength is about 10 mm above the desired wavelength and then scan back to the original point and stop.

6. To obtain a derivative spectra, bring the pen over to the chart paper and start the scan.

7. Superimpose the absorption spectra over the derivative spectra by recording the absorption spectra at the same scanning speed as was used for the derivative spectra.

8. Using a "blank" solution that contains only 9.5% HCl in ethanol, obtain a derivative and an absorption spectra and superimpose them on the spectras obtained for the standard solutions (use the same instrument settings).

Discussion

Derivative spectrometry is useful in identifying compounds with similar structures that absorb energy and give complex spectra. An example is found in phenprocoumon, warfarin, and acenocoumarol. These compounds give UV spectra that are highly similar, but their derivative spectra are very dissimilar. One method of comparison is through examination of the wavelengths at which the derivative spectra cross the derivative "blank" baseline. These wavelengths differ for the three compounds. Other maximum and minimum wavelengths in the derivative mode are also sufficiently different so that identification is aided.

Phenprocoumon[31] is a prothrombopenic anticoagulant used for several malfunctions, including pulmonary embolism, in order to prevent further embolisms. It is also used for the treatment of thrombophlebitis and injury to blood vessels. After phenprocoumon is taken, 48 to 72 hours are necessary for initial effectiveness. Phenprocoumon, as with other drugs, may have several unpleasant side effects. Overdose may cause hemorrhagic diathesis. Vitamin K can be used to overcome the overdose, but its analgesic activity is slow to take effect. Even when ingested in the prescribed amounts, the compound may cause diarrhea and dermatitis.

Phenprocoumon has the following structure:

The International Union of Pure and Applied Chemistry (IUPAC) name for warfarin is 3-α-phenyl-Ā-acetylethyl-4-hydroxycoumarin. The structure of warfarin is

Warfarin[32] is prepared as a powder. It is tasteless and odorless. Like other organic compounds of its type, it is soluble in nonpolar solvents such as ether, but insoluble in polar solvents such as water. However, it is also prepared as water-soluble sodium or potassium salt. Warfarin was first prepared at the University of Wisconsin as an outgrowth of research on the sweet clover disease of cattle, which is characterized by hemorrhaging. Hydroxycoumarin was found to be a constituent of this clover.

Warfarin is used as a coagulant and is used as a medication for thromboembolic malfunctions such as pulmonary embolism and coronary thrombosis. Warfarin inhibits prothrombin formation in the liver. Warfarin is also used as a rodenticide because it can induce extreme hemorrhaging leading to death when taken in large amounts.

Acenocoumaral[33] has the IUPAC name of 3-(α-acetenyl-*p*-nitrobenzyl)-hydroxycoumarin and its molecular formula is $C_{19}H_{15}NO_6$. Like the compounds previously mentioned, it is an anticoagulant that inhibits prothrombin formation in the liver.

Questions

1. *Besides derivative spectrometry, name three other methods by which organic compounds are identified.*

2. *Why do so many compounds have similar ultraviolet spectra?*

3. *When scanning speeds change, different derivative spectra are obtained. Why?*

4. *Name two or three other compounds that have similar UV spectra but that might possibly be identified using derivative spectroscopy.*

5. *Name several uses of the compounds acenocoumarol, phenprocoumon, and warfarin.*

6. *Suggest how derivative spectrometry may be used to quantitate the compounds discussed in this experiment.*

Answers to these questions may be found beginning on page 186.

EXPERIMENT 9
DERIVATIVE SPECTROMETRY ANALYSIS OF ACETAMINOPHEN AND SODIUM SALICYLATE IN TABLETS

Introduction

There are no official methods to analyze acetaminophen and sodium salicylate when combined into a single drug. Derivative spectrometry was thought to have the proper selectivity for the analysis of these drugs due to the 50 nm or more difference between their maximum absorbance wavelengths in most solvents as well as the fact that their UV absorption spectra overlap.

In derivative spectroscopy, the wavelength vs. the first or second derivative of spectral intensity or of absorbance is recorded. Derivative spectroscopy has been found to be helpful in quantitative analysis when a smaller peak is interfered with by some other large overlapping peak.

There are several different ways to obtain derivative spectra. Numerical methods include those in which the spectrum is recorded numerically and derivatives are obtained using a computer.[34] Another method is wavelength modulation. In this method, the selected wavelength is scanned quickly back and forth over a small spectral segment. This produces an alternating current in the photodetector. This current can be measured electronically. If the scanned spectral segment is small compared to the entire absorption band associated with it, then the alternating current produced by the wavelength modulation is very closely proportional to the first derivative with respect to wavelength. For example, visualize an absorption spectra of a compound that has zero absorbance at 250 nm, absorbs maximally at 300 nm, and has zero absorbance at 350 nm. When the wavelength near 255 nm is rapidly and repeatedly scanned back and forth, the phototube detector responds to give a spectra in which intensity is proportional to the slope of the absorption band. It is obvious that the slope (curvature or rate of change in intensity with wavelength) of the absorption band is greater at about 280 nm than it is at 250 nm, and, therefore, because of this greater change in intensity, a more pronounced first derivative spectrum is obtained.

Wavelength modulation instruments use a frequency- and phase-selective alternating current (AC) amplification system to measure the intensity change. In order to obtain the first derivative, the AC amplification device is set to the frequency of wavelength modulation. To produce the second derivative, the AC amplification devices are set on twice the frequency of wavelength modulation. Both sinusoidal and square wave modulation waveform may be used.[35]

There are several ways in which wavelength modulation can be brought about: use of a vibrating mirror so that light of two different wavelengths can be seen by the phototube; oscillation of an interference filter; placement of a rotating refractor plate in the light beam inside the monochromator; and oscillation of the slits, mirror, grating, or prism of the monochromator.

Electronic differentiators are used to produce time derivatives of the absorption spectra.[36]

Objectives

The objectives of this experiment are to become familiar with the derivative spectrometer, and to show how derivative spectroscopy can be used to quantitatively analyze for certain organic compounds.

Instrumentation, Solutions, and Reagents

Obtain a Cary Model 118 spectrophotometer or equivalent with derivative attachment. (See Section IV for other derivative spectrophotometers.) Prepare 4 N alcoholic HCl in the following way:

1. Pipette 34.0 ml of concentrated HCl into a 1000-ml volumetric flask and dilute to 1000 ml with 95% ethanol.
2. Mix the solution well

Obtain the standard compounds acetaminophen and sodium salicylate (these compounds can be ordered from the USP. Prepare acetaminophen standard solution:

1. Weigh 36 mg of acetaminophen into a 100-ml volumetric flask and dilute to 100 ml with 95% ethanol to obtain a 360-µg/ml solution.
2. Pipette 5.0 ml of the 360-µg/ml solution into a 100-ml volumetric flask.
   - Add 4.5 ml distilled water, 40 ml 95% ethanol, and 5 ml of alcoholic HCl.
   - Dilute to 100 ml with ethanol and mix well to obtain an 18-µg/ml solution.

Next, prepare sodium salicylate standard solution:

1. Weigh 60 mg of sodium salicylate into a 100-ml volumetric flask and dilute to 100 ml with 95% ethanol to obtain a 600-µg/ml solution.
2. Pipette 5.0 ml of the 600-µg/ml solution into a 100-ml volumetric flask.
   - Add 4.5 ml distilled water, 40 ml 95% ethanol, and 5 ml of alcoholic HCl.
   - Dilute to 100 ml with ethanol and mix well to obtain a 30-µg solution.

Prepare a mixed standard stock solution containing an acetaminophen concentration of 360 µg/ml and sodium salicylate at 600 µg/ml:

1. Weigh 36 mg of acetaminophen and 60 mg of sodium salicylate into the same 100-ml volumetric flask.
2. Dilute to 100 ml with 95% ethanol and mix well.

Prepare a mixed working standard solution in the following way:

1. Pipette 4.5 ml distilled water into four 100-ml volumetric flasks.
2. Pipette 0-, 4.0-, 5.0-, and 6.0-ml aliquots, respectively, of the standard stock solution into the four volumetric flasks.
3. To each flask, add 40 ml ethanol followed by 5 ml of 4 N alcoholic HCl.
4. Dilute to 100 ml with ethanol.
5. Calculate the concentration of each standard solution in the following way:

(ml stock solution) × 360 µg acetaminophen/ml × 1/100 ml
　　　　　　　　　　= conc. of acetaminophen (µg/ml)

(ml stock solution) × 600 µg sodium salicylate/ml × 1/100 ml
　　　　　　　　　　= conc. of sodium salicylate (µg/ml)

For example, for 4.0 ml of stock solution:

$$4 \text{ ml} \times 360 \text{ } \mu g \text{ acetaminophen/ml} \times 1/100 \text{ ml} = 14.4 \text{ } \mu g/ml$$

$$4 \text{ ml} \times 600 \text{ } \mu g \text{ sodium salicylate/ml} \times 1/100 \text{ ml} = 24 \text{ } \mu g/ml$$

Commercial tablets usually contain 180 mg acetaminophen and 300 mg sodium salicylate per tablet.

- For the purpose of demonstrating the utility of the analysis, weigh 180 mg of acetaminophen and 300 mg sodium salicylate into the same 500-ml volumetric flask.

- Dilute the acetaminophen and sodium salicylate to 500 ml with ethanol and mix well to obtain approximately 360 $\mu g/ml$ acetaminophen and 600 $\mu g/ml$ sodium salicylate. This will be called the "sample".

Pipette 5.0 ml of the sample solution into a 100-ml volumetric flask.

1. Add 4.5 ml of water followed by 40 ml of ethanol and 5 ml of 4 *N* alcoholic HCl.

2. Dilute to 100 ml with ethanol and mix well.

Procedure

Take each of the single standard solutions of acetaminophen and sodium salicylate and treat as follows using the Cary 118 spectrophotometer with derivative attachment (or an equivalent spectrophotometer). (The directions given are based on the assumption that the Cary 118 is being used.)

1. Scan and record an absorption spectra following the manufacturer's manual. Start at 360 nm and scan to 200 nm.

2. Return the scan to the original starting wavelength, but use a fresh portion of the chart paper.

3. Set the function switch to "Deriv." Lift the pen so it is held away from the chart paper.

4. Try different settings of the "Abs. Range" switch and "Scan Speed Selector" and run scans. Choose settings that will result in derivative peaks that are about 50% full-scale deflection and will produce the maximum separation of the first derivative peaks of acetaminophen and sodium salicylate. The settings of abs. range and scan speed that are finally selected must be the same for both acetaminophen and sodium salicylate.

5. Scan back until the wavelength is about 10 nm above the starting wavelength. (*Note:* the beginning and ending wavelength will be the same as that used for the absorption spectra.) Scan to the starting wavelength and stop the scan.

6. Place the tip of the pen on the chart paper and start the scan to obtain a derivative spectra for each of the single standards. Use different

areas of the chart paper for the various standards. On the same area of the chart paper, obtain the derivative spectra of acetaminophen superimposed over the derivative spectra of sodium salicylate. Also obtain the derivative spectra of the "blank" on this same part of the chart paper.

7. Examine the derivative spectra of acetaminophen superimposed over sodium salicylate and pick the wavelength(s) in which peak heights should be measured for quantitation of these compounds. The ideal wavelength would be one in which the derivative peak of one of the compounds is at the baseline drawn by the blank, while the derivative peak of the other compound is at its maxima. For acetaminophen, the peak to be measured is nearly 261 nm, while for sodium salicylate it is nearly 317 nm. The peaks are measured from the maxima down to the baseline produced with the "blank".

8. Using the same chart area, scanning speed, abs. range, and other parameters found in the preceding steps, obtain the derivative spectra of the standards containing 0, 4.0, 5.0, and 6.0 ml of the stock standard solution in 100 ml of acidified ethanol.

9. Using the same parameters as used in the preceding step, obtain the derivative spectra of the "sample" and the "blank" in a different area of the chart.

10. Prepare a standard curve for acetaminophen and salicylic acid by plotting concentration vs. peak height using the working standards containing 0, 4.0, 5.0, and 6.0 ml stock standard solution in 100 ml of acidified ethanol.

11. Using the standard curve, determine the concentration of acetaminophen and sodium salicylate in the sample in micrograms per milliliter.

12. Calculate the percentage of recovery of acetaminophen and sodium salicylate in the sample using the following equations:

$$\% \text{ Recovery of Acetaminophen} = \frac{\mu g/ml \text{ Acetaminophen found}}{\mu g/ml \text{ Acetamiophen prepared}} \times 100$$

$$\% \text{ Recovery Sodium Salicylate} = \frac{\mu g/ml \text{ Sodium Salicylate found}}{\mu g/ml \text{ Sodium Salicylate added}} \times 100$$

Discussion

This experiment illustrates how derivative spectrometry can be used to quantitate, without separation, two compounds that have overlapping absorption spectra.

Acetaminophen has the IUPAC name of *N*-(4 hydroxyphenyl)-4' hydroxyacetanilide. Its structural formula is[37]

$$HO-\!\!\bigcirc\!\!-NHCOCH_3$$

Acetaminophen is a nonaddicting analgesic and antipyretic. Analgesics are substances that alleviate pain by elevating the pain threshold value (the point at which pain is felt). This threshold value is a function of sex, age, circulatory change, skin temperature, fear, emotion, and other factors and is not consistent in the same individual or between individuals. An antipyretic is a substance that reduces body temperature. Exactly how analgesics work is not completely understood. It is believed that they act by interfering with pain signals from sensory nerve tracts in the subcortical levels of the brain.

Many analgesics are also antipyretics. There are three general classes of analgesics: (1) nonopiate and nonaddicting analgesics and antipyretics, (2) opiate analgesics, and (3) nonopiate addicting analgesics.

Sodium salicylate has the structural formula

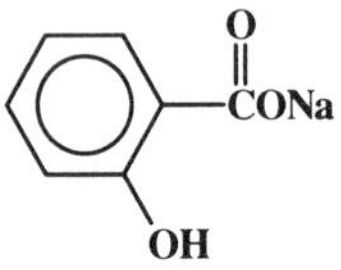

$$\text{C}_6\text{H}_4(\text{OH})\text{CONa} \text{(with C=O)}$$

It occurs as a microcrystalline powder or as scales.[38] It is either odorless or has a faint odor and is affected by light. Sodium salicylate is an analgesic and antipyretic used to alleviate pain and to reduce temperature. For example, it can be used to alleviate the symptoms of gout or rheumatic fever.

Questions

1. *Describe methods other than derivative spectrometry in which mixtures of sodium salicylate and acetaminophen can be analyzed.*

2. *Could derivative fluorescence be used to analyze mixtures of acetaminophen and sodium salicylate? Why or why not?*

3. *Name several mixtures of compounds that you think could not be analyzed by derivative spectrometry and explain why.*

4. *Using a Cary 118 spectrophotometric manual or some other manufacturer's derivative spectrometric manual, draw a schematic of a derivative spectrometer. Label all the parts and give their function.*

5. *Assume you have a solution containing 36 mg acetaminophen and 60 mg of sodium salicylate in 100 ml of 95% ethanol. Of this, 5.0 ml is taken and diluted to 50.0 ml with 95% ethanol. Of this solution, 5.0 ml is taken and diluted to 100.0 ml with 95% ethanol. What is the final concentration of acetaminophen and sodium salicylate in micrograms per milliliter?*

6. *Tablets that are given to patients contain ingredients other than acetaminophen and sodium salicylate, such as diluents, colors, lubricants, disintegrants, starches, and lactose. Many of these ingredients are insoluble in ethanol. When tablets are analyzed by the derivative method described, what steps should probably be taken before the derivative spectrum of the sample is secured?*

Answers to these questions may be found beginning on page 188.

EXPERIMENT 10
ANALYSIS OF PHENYTOIN SODIUM USING DERIVATIVE SPECTROMETRY

Introduction

Phenytoin sodium is difficult to analyze by UV absorption techniques in water because the UV absorption spectra of the drug in water do not show distinct absorption maxima. Thus derivative spectrometry is a desirable alternative quantitative method.

Several different commercial spectrometers can produce a graphic display of the first and second derivative of the analog signal given by the spectrophotometer. The first derivative can be thought of as the change in absorbance (dA) with change in wavelength (dλ). The derivative technique can allow resolution of two spectral lines that have very close wavelengths. In addition, derivative spectrometry can be used for absorption spectra that do not have pronounced maximas. The latter condition is the case with phenytoin sodium which shows a very pronounced derivative maxima that can be utilized for quantitative analysis.

Objectives

The objectives of Experiment 10 are to show how derivative spectroscopy can be used to analyze compounds without distinct absorption bands, and to increase familiarity with derivative spectroscopy techniques.

Instrumentation, Solutions, and Reagents

Obtain a Cary Model 118 spectrophotometer or equivalent with derivative attachment. (Section IV lists other derivative spectrometers.) Obtain distilled or deionized water and secure phenytoin sodium standard from the USP. Prepare phenytoin sodium standard solution:

1. Weigh 11.1 mg of USP phenytoin sodium standard into a 100-ml volumetric flask.
2. Dilute to 100 ml with distilled or deionized water to obtain a 110-μg/ml solution.

Commercial capsules contain varying amounts (30, 100, and 250 mg) of phenytoin sodium. For the purpose of indicating how the derivative method can be used for analysis, weigh 22 mg of phenytoin sodium standard into a 200-ml volumetric flask and dilute to 200 ml with distilled or deionized water and mix well. This solution will be called the sample.

Procedure

Take the standard solution of phenytoin sodium in water and handle in the way described below using the Cary Model 118 spectrometer with derivative attachment or equivalent. (The instructions given are based on use of the Cary 118 spectrometer. If another instrument is used, follow manufacturer's directions.)

1. Scan and record an absorption spectra of wavelengths between 250 and 320 nm (follow the manufacturer's instructions).
2. Return to the original wavelength, but use an untouched part of the chart paper.

3. Set the function switch to "Deriv." and lift the pen away from the chart paper.

4. Try different positions of the "Abs. Range" switch and run scans between 250 and 320 nm. Choose settings that produce derivative peaks that are about 50% full-scale deflection.

5. Scan up until the wavelength is about 5 nm above the beginning wavelength, then scan back to the starting wavelength and stop. (In obtaining absorption spectra, the usual procedure is to start at the higher wavelength and scan down to the lower wavelength.)

6. Set the pen on the chart paper and begin the scan to obtain a derivative spectra of the standard.

7. In the same area of the chart paper used with the standard, obtain a derivative spectra of a "blank" solution of distilled or deionized water.

8. Using the same instrumental parameters, obtain the derivative spectra of the "sample".

9. Using the derivative peak chosen earlier, obtain the peak height by measuring from the "blank" spectra up to the maximum peak.

10. Calculate the concentration of phenytoin sodium in the sample using the following formula:

$$\begin{array}{cccc} \text{Phenytoin} & = \text{Peak height} & \times \text{ Conc. standard} & \times \text{ Final volume} \\ \text{Na}(\mu g) \text{ in sample} & \text{(sample)} & (\mu g/ml) & \text{(sample)} \end{array}$$

11. Calculate the percentage of recovery of phenytoin sodium in the sample using the following equation:

$$\% \text{ Recovery phenytoin Na} = \frac{\mu g \text{ Phenytoin Na found}}{\mu g \text{ Phenytoin Na prepared}} \times 100$$

Discussion

This experiment shows how derivative spectroscopy can be used to analyze compounds that do not have pronounced UV absorption maxima. Several drugs could best be analyzed by this method.

Phenytoin sodium has the molecular formula $C_{15}H_4N_2NaO_2$. The IUPAC name is 2,4-imidazolinedione,5,5-diphenyl-monosodium salt.[39] Phenytoin sodium has the following structural formula[40]

Phenytoin sodium is used by the medical profession to treat various illnesses that manifest themselves by involuntary movement:[41] epilepsy, chorea, and Parkinson's syndrome. It has also been used for the modification of premature ventricular contractions and paroxysmal ventricular and supraventricular tachycardias. Various other uses include treatment of migraine headaches, trigeminal neuralgia, and even some psychoses.

Questions

1. Name at least one other compound not having a pronounced UV maxima that could be analyzed using derivative spectroscopy.

2. Why does the derivative of the phenytoin sodium UV spectra give distinct maxima while the UV absorption spectra do not?

3. The first derivative peak height of a sample of an impure phenytoin sodium was 25 mm at 258 nm while the corresponding peak height of a standard solution was 24 mm. If the standard concentration was 100 μg/ml and the sample consisted of 1.5 g of impure phenytoin sodium diluted to 100 ml in distilled water, calculate the concentration of phenytoin sodium in the sample and the percentage of phenytoin sodium in the sample.

4. A student mistakenly weighed 0.25 g of phenytoin sodium into a 10-ml volumetric flask and diluted to 10 ml with water. What was the concentration in micrograms per milliliter? How would the student take this solution and prepare a 100-μg/ml solution in water?

> **5.** *Phenytoin and phenytoin sodium both give similar first derivative spectra. If you had only phenytoin standard and you wished to analyze for phenytoin sodium, how would you accomplish this?*

Answers to these questions may be found beginning on page 190.

EXPERIMENT 11
DIRECT DERIVATIVE SPECTROSCOPY OF DRUGS IN THE PRESENCE OF INTERFERING EXCIPIENTS

Introduction

The ideal condition under which certain drugs and natural products can be analyzed by UV spectrometry would be one wherein the sample is simply diluted with the proper solvent and the UV absorption spectra obtained and compared with the UV-absorption spectra of a standard solution. However, when the sample contains UV-absorbing degradation products or added substances, the spectral overlapping of these absorption bands can serve to hide the absorption of the compound being analyzed and, thus, prevent its quantitation.

The U.S. Government allows suitable substances such as bases, carriers, coatings, colors, flavors, preservatives, stabilizers, and vehicles to be added to a pharmacopeial dosage form or finished device to enhance its stability, usefulness, or elegance or to facilitate its preparation.[42] Such substances are thought to be suitable if they are harmless in the amounts used, if they do not surpass the minimum amount required for their intended effect, if they do not hurt or limit the bioavailability or the therapeutic effect of the drug, and if they do not interfere with the analysis and qualitative tests that are used to determine whether the drug meets government standards.

The diluents, colors, lubricants, disintegrants, and adhesives referred to are usually starches, lactose, sucrose, or other nonharmful products. Tablets may be coated with various materials.

All of the components mentioned above may cause spectral interference. However, derivative spectroscopy allows resolution of two or more spectral bands of close wavelength, and thus allows quantitation without separation.

Objectives

The objectives of Experiment 11 are to illustrate how derivative spectroscopy can be used to analyze a commercial drug in the presence of interfering excipients, to compare results obtained from official methods of analysis, and to learn more about substances added to commonly used drugs.

Instrumentation, Solutions, and Reagents

Obtain a Cary Model 118 spectrophotometer or equivalent with derivative attachment. (Other derivative spectrophotometers are listed in Section IV.)

Obtain edrophium chloride standard from the USP.

Prepare 1 N HCl to be used for derivative analysis by taking 43 ml of concentrated HCl into a 500-ml graduate and dilute to 500 ml with water and mix well.

Prepare the following reagents for the USP analysis.[43]

1. 0.1 N HCl: place 25 ml of the 1 N HCl made earlier into a 250-ml graduate and dilute to 250 ml with water and mix well.

2. pH 8 phosphate buffer:
   - Weigh 27.22 g of monobasic potassium phosphate (KH_2PO_4) into a 1000-ml graduate and add distilled water to 1000 ml and mix well.
   - Weigh 4 g NaOH into a 500-ml graduate and dilute to 500 ml with water. Mix well to obtain 0.2 M NaOH.
   - Place 46.1 ml of 0.2 M NaOH and 50 ml of 0.2 M potassium phosphate monobasic into a 200-ml volumetric flask and dilute to 200 ml with distilled water. Mix well.

Obtain reagent-grade sodium chloride, hexane, and ethyl ether.

Obtain edrophonium chloride injection from a suitable pharmaceutical source and prepare edrophonium chloride standard solution for derivative use as follows:

1. Weigh 10 mg standard edrophonium chloride into a 50-ml volumetric flask and dilute to 50.0 ml with distilled water.

2. Place an aliquot equivalent to about 2 mg edrophonium chloride into a 50-ml volumetric flask and dilute to 50.0 ml with 1 N HCl. This solution is 40 μg/ml.

Prepare a standard solution of edrophonium chloride to be used for the USP analysis[43] using the following procedure:

1. Weigh 25 mg edrophonium chloride into a 100-ml volumetric flask and dilute to 100 ml with 0.1 *N* HCl to obtain a solution of 250 μg/ml.

2. Place an aliquot equivalent to 2500 μg into a 50-ml volumetric flask and dilute with 0.1 *N* HCl to obtain a concentration of 50 μg/ml.

Prepare a sample solution to be used for derivative analysis:

1. Place a volume of the injection equivalent to about 50 mg edrophonium chloride into a 250-ml volumetric flask and dilute to 250 ml with distilled water, mixing well.

2. Place an aliquot equivalent to about 2 mg into a 50-ml volumetric flask and dilute to 50 ml with 1 *N* HCl and mix well. This will contain about 40 μg/ml edrophonium chloride.

Prepare the sample solution to be used in the USP analysis by pipetting a volume of the edrophonium chloride injection equivalent to about 50 mg of the active ingredient into a glass-stoppered 50-ml centrifuge tube.

Procedure

For the derivative analysis, take the standard solution of edrophonium chloride prepared and treat as follows using the Cary Model 118 spectrophotometer with derivative attachment or equivalent. The instructions given are based on use of the Cary Model 118 spectrophotometer.

1. Scan and record an absorption spectra between the wavelengths 350 to 260 nm by following the manufacturer's instructions.

2. Return to the original wavelength and use an untouched portion of the chart paper.

3. Set the function switch to "Deriv." and lift the pen away from the paper.

4. Try different positions of the "Abs. Range" switch and run scans between 350 and 260 nm. Choose settings that produce derivative peaks that possess about 50% full-scale deflection. Pick an appropriate peak to use for quantitation.

5. Scan back until the wavelength is about 5 nm above the beginning wavelength and then scan slowly down to the beginning wavelength and stop. (Normal practice in absorption spectrometry is to scan from higher to lower wavelengths.)

6. Set the pen on the chart paper and begin the scan in order to obtain a derivative spectra of the standard.

In the same area of the chart paper used with the standard, obtain a derivative spectra of a "blank" solution made up of 10.0 ml of distilled water diluted to 50 ml with 1 *N* HCl in a volumetric flask. Using the same operating settings, obtain the derivative spectra of the sample.

Using the derivative peak chosen earlier, obtain the peak height by measuring from the "blank" spectra up to the maximum peak.

Calculate the concentration of edrophonium chloride in the sample using the following formulas:

$$\begin{bmatrix} \text{Edrophonium} \\ \text{chloride in sample (mg/ml)} \end{bmatrix}$$

$$= \begin{bmatrix} \text{Net peak} \\ \text{height (sample)} \end{bmatrix} \times \begin{bmatrix} \text{Conc.} \\ \text{standard } (\mu\text{g/ml}) \end{bmatrix} \times \begin{bmatrix} \dfrac{1 \text{ mg}}{1000 \text{ } \mu\text{g}} \end{bmatrix} \times \begin{bmatrix} \text{Dilution} \\ \text{factor} \end{bmatrix}$$

$$\begin{bmatrix} \text{Edrophonium} \\ \text{chloride in sample (mg/ml)} \end{bmatrix}$$

$$= \begin{bmatrix} \text{Net peak} \\ \text{height (sample)} \end{bmatrix} \times \begin{bmatrix} \text{Conc.} \\ \text{standard } (\mu\text{g/ml}) \end{bmatrix} \times \begin{bmatrix} \dfrac{250 \text{ ml}}{10 \text{ ml}} \\ \text{Aliquot} \end{bmatrix} \times \begin{bmatrix} \dfrac{50 \text{ ml}}{\text{Sample}} \\ \text{taken (ml)} \end{bmatrix}$$

$$\begin{bmatrix} \text{Declared edrophonium} \\ \text{chloride } (\%) \end{bmatrix}$$

$$= \begin{bmatrix} \text{Edrophonium} \\ \text{chloride found } (\mu\text{g/ml}) \end{bmatrix} \Big/ \begin{bmatrix} \text{Edrophonium chloride} \\ \text{declared } (\mu\text{g/ml}) \end{bmatrix} \times 100$$

For the USP analysis, take a sample solution of edrophonium chloride in a 50-ml glass-stoppered centrifuge tube and treat in the following way:

1. Add 5 ml of pH 8.0 phosphate buffer and 5 g of sodium chloride.

2. Wash the solution with four 20-ml portions of a mixture of 1:1 hexane-ethyl ether using these procedures:

   - add 20 ml 1:1 hexane-ethyl ether
   - carefully shake about 30 s
   - centrifuge about 1 min at high speed
   - use a disposable pipette to siphon off the top hexane-ether layer and discard

3. Transfer the aqueous phase to a 100-ml volumetric flask, add 0.1 N HCl, and mix.

4. Transfer a 5-ml aliquot of this solution to a 50-ml volumetric flask, add 0.1 N HCl to volume, and mix.

5. Concomitantly determine the absorbances of sample and standard soluution (50 μg/ml) in 1-cm cells at the wavelength of maximum absorbance (about 275 nm) with a suitable spectrophotometer using 0.1 N HCl as the blank.

6. Calculate the quantity of the edrophonium chloride in milligrams per milliliter of the injection taken by the formula given below:

$$\begin{bmatrix} \text{Edrophonium} \\ \text{chloride (mg/ml)} \end{bmatrix}$$

$$= \begin{bmatrix} \dfrac{\text{Net absorbance,}}{\text{sample}} \\ \dfrac{}{\text{Net absorbance,}} \\ \text{standard} \end{bmatrix} \times \begin{bmatrix} \text{Conc.} \\ \text{standard } (\mu\text{g/ml}) \end{bmatrix} \times \begin{bmatrix} \text{Dilution} \\ \text{factor} \end{bmatrix} \times \begin{bmatrix} \dfrac{\text{Final vol.}}{\text{(ml) sample}} \\ \dfrac{}{\text{Original}} \\ \text{sample taken (ml)} \end{bmatrix}$$

$$\begin{bmatrix} \text{Edrophonium} \\ \text{chloride (mg/ml)} \end{bmatrix}$$

$$= \begin{bmatrix} \dfrac{\text{Net absorbance,}}{\text{sample}} \\ \dfrac{}{\text{Net absorbance,}} \\ \text{standard} \end{bmatrix} \times \begin{bmatrix} \text{Conc.} \\ \text{Std. } (\mu\text{g/ml}) \end{bmatrix} \times \begin{bmatrix} \dfrac{1 \text{ mg}}{1000 \ \mu\text{g}} \end{bmatrix} \times \begin{bmatrix} \dfrac{100 \text{ ml}}{\text{Aliquot (5 ml)}} \end{bmatrix} \times \begin{bmatrix} \dfrac{\text{Sample (vol)}}{\text{Original}} \\ \text{sample taken (ml)} \end{bmatrix}$$

7. Calculate the percent declared as follows:

$$\begin{bmatrix} \text{Declared edrophonium} \\ \text{chloride } (\%) \end{bmatrix}$$

$$= \begin{bmatrix} \text{Edrophonium chloride} \\ \text{found (mg/ml)} \end{bmatrix} \Big/ \begin{bmatrix} \text{Edrophonium chloride} \\ \text{declared (mg/ml)} \end{bmatrix} \times 100$$

Compare the edrophonium chloride (mg/ml) found through derivative analysis to that found by the USP analysis.

Discussion

The edrophonium chloride injection is an aqueous solution buffered to a pH of about 5.4 with a citrate buffer system. Some preparations also contain phenol as a preservative. These excipients interfere with the direct spectrophotometric analysis of this drug without separation. Using derivative spectrometry, it has been shown that the drug can be quantitated in the presence of excipients.

Edrophonium chloride has the following structure:

Edrophonium chloride is a cholinomimetic drug with prominent nerve-muscular action and weak ganglionic and cholinergic neuroeffector stimulant action. As an anticurare agent[44] the transmission of an impulse from a nerve to the skeletal muscle fibers that it enervates is inhibited at the neuromuscular junction.[45] Curarimimetic agents act to depress respiration. Edrophonium can act to reverse the effect of the curarimimetic agents. Cholinergic agents increase the ability of acetylcholine, liberated at parasympathetic nerve endings, to activate the tissue that it enervates.[46] Cholinergic agents are parasympathomimetic agents and cholinesterase inhibitors.[47] A parasympathomimetic agent is one that effects visceral organ response qualitatively similar to that produced by simulating the parasympathetic innervation of such tissue. Parasympathetic innervation refers to nerves connected to involuntary organs. A cholinesterase enzyme inhibitor produces a parasympathetic effect by inhibiting the rapid destruction of the cholinergic mediator of nerve impulses, acetylcholine, at the neuromuscular junction.

Questions

1. *What are the advantages of being able to analyze for a drug without separation of interfering excipients?*

2. *Name three other drug preparations for possible analysis by derivative spectroscopy without separation from excipients.*

3. *What methods other than derivative spectroscopy have been used to correct for background interference?*

4. *Using the USP method to analyze for edrophonium chloride, the net peak height for the sample was 75 mm, while the net peak height of the standard was 74 mm. Assuming a standard concentration of 55 μg/ml, calculate the milligrams per milliliter found in the edrophonium chloride injection if a 10-ml sample is taken for analysis.*

5. *Using the derivative analysis described in this experiment, calculate the milligrams per milliliter of edrophonium chloride in the injection if given the following data: net peak height sample, 50 mm; net peak height sample, 49 mm; concentration standard; 44 µg/ml; ml original sample taken for analysis, 10 ml; and aliquot taken from 250-ml sample solution, 10 ml.*

Answers to these questions may be found beginning on page 192.

EXPERIMENT 12
ANALYSIS OF 6α-METHYL TESTOSTERONE IN THE PRESENCE OF 6β-METHYL TESTOSTERONE

Introduction

The analysis of compounds with overlapping spectral bands presents some difficulty because it is not always possible to locate these bands. The use of derivative spectroscopy can allow the separation and inspection of these bands in a large number of cases. The wavelengths of maxima and minima in the derivative spectra probably will not exactly match the wavelength of the minima and maxima, respectively, of the absorption spectra because the derivatives are those of composite absorption spectra. When a derivative spectra is compared to the absorption spectra of the same compound, it is found that each maxima in the derivative curve coincides with the point of maximum slope on the corresponding shoulder of the UV curve, and the derivative minima and zero values represent the true location of the flexes and maxima in the UV absorption spectra. The derivative curve recorded is a function of the change in absorbance with change in wavelength $(dA/dk\lambda)$, and thus the height of the derivative spectra is not only determined by concentration but also by rate of wavelength scan.

Objective

The sole objective of this experiment is to analyze a typical steroid using derivative techniques.

Instrumentation, Solution, and Reagents

Obtain a Cary Model 118 spectrophotometer (or equivalent) with derivative attachment. (Section IV lists other derivative spectrophotometers.)

Obtain 6α-methyl testosterone and 6β-methyl testosterone from the USP.

Obtain diethylene glycol dimethyl ether solvent and prepare 6α-methyl testosterone standard solution:

1. Weigh 25 mg standard 6α-methyl testosterone into a 10-ml flask.
2. Pipette 5.0 ml diethylene glycol dimethyl ether into the same 10-ml volumetric flask and mix well to obtain a 5-mg/ml solution.

Prepare 6β-methyl testosterone standard solution:

1. Weigh 25 mg standard 6β-methyl testosterone standard into a 10-ml volumetric flask.
2. Pipette 5.0 ml diethylene glycol dimethyl ether into the same 10-ml flask and mix well to obtain a 5-mg/ml solution.

Prepare a mixed standard solution:

1. Weigh 25 mg 6α-methyl testosterone and 25 mg of 6β-methyl testosterone into the same 10-ml flask.
2. Add 5.0 ml diethylene glycol dimethyl ether into the 10-ml volumetric flask and mix well to obtain a 5-mg/ml concentration of each standard.

Commercial steroid tablets contain various amounts of steroids. For the purpose of demonstrating the utility of derivative spectroscopy in this type of analysis, the following sample is prepared:

1. Weigh 24.5 mg 6α-methyl testosterone and 24.6 mg 6β-methyl testosterone into the same 10-ml volumetric flask.
2. Add 5.0 ml diethyleneglycol dimethyl ether and mix well. This will be the ''sample''.

Procedure

Take each of the single standard solutions of 6α- and 6β-methyl testosterone and follow the directions given below, using the Cary Model 118 spectrophotometer with derivative attachment. (An equivalent derivative spectrometer can be used, but the directions given here assume the Cary Model 118 is being used.)

1. Scan and record an absorption spectra of each solution on a separate portion of the chart paper. Follow the directions given in the manufacturer's manual. Start at 380 nm and scan back to 260 nm.
2. Return the scan to the original starting wavelength, but use an untouched portion of the chart paper.
3. Set the function switch to ''Deriv.'' Lift the pen away from the paper.
4. Try different settings of the ''Abs. Range'' switch and ''Scan Speed'' selector and run scans. Choose settings that will result in derivative

peaks that are about 50% full-scale deflection and will produce the maximum separation of the first derivative peak of 6α- and 6β-methyl testosterone at about 300 nm. The settings selected must be the same for both 6α- and 6β-methyl testosterone.

5. Scan back until the wavelength is about 10 nm above the beginning wavelength, which is 380 nm, and then scan to this point and stop.

6. Place the pen point on the chart paper and start the scan to obtain a derivative spectra for each of the single standards in different areas of the chart paper.

On an unused portion of the chart paper, using the same parameters utilized for the single standards, obtain a derivative spectra of the mixed standard solution of 6α- and 6β-methyl testosterone. In this same area, obtain the derivative spectra of a "blank" solution of diethylene glycol dimethyl ether.

Using the same parameters as used in the preceding step, in a different area of the chart paper, obtain the derivative spectra of the sample and the blank.

Obtain the net peak height of the 6α-methyl testosterone at about 300 nm of the sample and the mixed standard by measuring the difference between the blank and the sample or standard peaks.

Calculate the quantity of 6α-methyl testosterone in the sample in the following way:

$$\begin{bmatrix} 6\alpha\text{-Methyl testosterone} \\ \text{(mg/ml in sample)} \end{bmatrix}$$
$$= \begin{bmatrix} \text{Net peak} \\ \text{height sample} \end{bmatrix} \Big/ \begin{bmatrix} \text{Net peak} \\ \text{height standard} \end{bmatrix} \times \begin{bmatrix} \text{Conc. std.} \\ \text{(mg/ml)} \end{bmatrix}$$

Calculate the percentage of recovery in the following manner:

% Recovery α-methyl testosterone in sample = (mg/ml)6α-methyl testosterone

recovered/6α-methyl testosterone added (mg/ml) × 100

Discussion

Testosterone has the following structure:[48]

The IUPAC name is Δ^4-androstene-17β-al-3-one. Based on this structure, 6β-methyl testosterone and the above compound are related. 6α-methyl testosterone has the following structure:

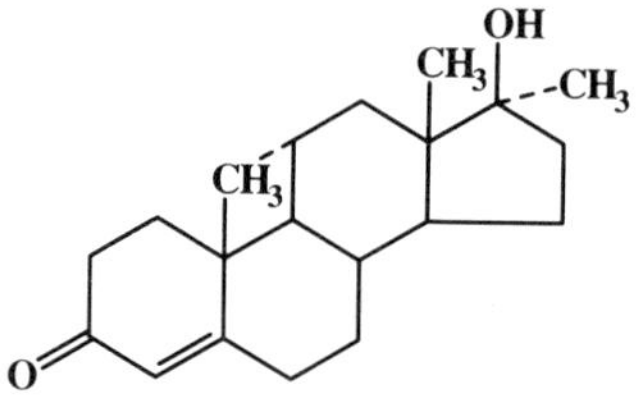

The dotted lines indicate the α position.

Another steroid, progesterone, has the following structure:[49]

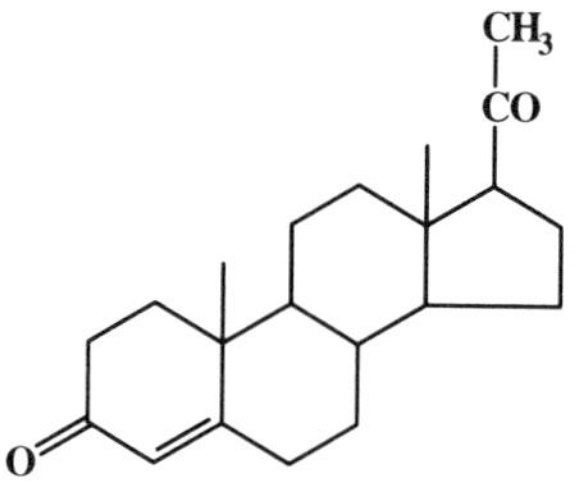

The British Pharmacopeia (BP) has named progesterone as 4-pregnene 3:20 dione.[50]

Testosterone is considered to be the male sex hormone. It is secreted by the testis and is responsible for the secondary sex characteristics as well as the development of the accessory sexual organs of the male. Testosterone has been found to be useful in overcoming the side effects of castration. It can cause premature virilism in the young male and in females. Testosterone also acts to inhibit the function of the pituitary gland. Testosterone has many therapeutic uses:[51] treatment for eunuchism, hyponadism or deficiency of the testis, bilateral interabdominal undescended testis, pituitary dwarfism, the male climacteric, impotence from glandular causes, Cushing's syndrome, Addison's disease, and hypopituitarism.

Castration of a young male causes the loss of many of the secondary male characteristics. The castrated male experiences ''hot flashes'' in which he feels that blood is rapidly going to his head and he perspires. His voice may become higher and more feminine. Sexual desire ceases. His beard grows more slowly and he may lose other body hair. When testosterone in the form of testosterone propionate is administered, these conditions disappear, but he must remain on the drug permanently.

Questions

1. *Name one other steroid mixture that could be analyzed using derivative spectrometry.*

2. *In your opinion, what causes the intense maximum at about 300 nm in the derivative spectra of the 6α-methyl testosterone compounds that is not found in the 6β-methyl testosterone?*

3. *Name two other steroids (besides testosterone) found in humans and their uses.*

4. *If you had a mixture of the two steroids mentioned in Question 3, propose a possible derivative spectrophotometric analysis of one of them in the presence of the other.*

5. *If a mixture of 6α- and 6β-methyl testosterone gave a net derivative peak of 25 nm at 300 nm, while a corresponding standard mixture containing 5.2 µg/ml 6α-methyl testosterone and 4.5 mg/ml 6β-methyl testosterone gave a net peak height of 24 nm for 6α-methyl testosterone under the same circumstances, calculate the concentration of 6α-methyl testosterone in the sample. Assume similar volumes of the sample and the standard.*

Answers to these questions may be found beginning on page 194.

EXPERIMENT 13
ANALYSIS OF SACCHARIN IN COMMERCIAL DRINKS. COMPARISON ANALYSIS BY FIRST AND SECOND DERIVATIVES

Introduction

Derivative spectrometry can be used to enrich spectral data. The strength of narrow absorbance bands increases compared to that of broad bands. Thus, strongly marked spectral bands of low strength may be systematically noted even if they are situated within a broad, intense band. Derivative spectrometry has been used to analyze many types of compounds, but the use of an unsuitable derivative method may lead to an analytical error. Picking the correct derivative method is a function of the compounds involved as well as spectral interferences present.

Objectives

The objectives of Experiment 13 are to analyze an artificial sweetener, and to compare the analysis made with the first derivative to the analysis made with the second derivative.

Instrumentation, Solutions, and Reagents

Obtain a Cary Model 118 spectrophotometer or equivalent with derivative attachment (see also Section IV).

Prepare 1 N NaOH by weighing 4 g NaOH into a 100-ml graduate and dilute to 100 ml with distilled water and mix well.

Obtain standard saccharin from the USP.

Prepare a 1000 ppm stock standard solution by weighing 100 mg of saccharin into a 100-ml volumetric flask and dilute to 100 ml with distilled water and mix well. This gives 1000 ppm as shown: 1000 μg/ml = 1000 μg/g = 1000 ppm.

Prepare working standard solutions:

1. Pipette 1 ml of the 1000-μg/ml solution into a 10-ml volumetric flask. Dilute to 10 ml with distilled water and mix well to obtain 100 ppm.

2. Pipette 0, 3, 6, and 9 ml of the 100-μg/ml solution into separate volumetric flasks, dilute to 100 ml with water, and mix well to obtain 3-, 6-, and 9-ppm solutions.

Find a cola drink that has no saccharin or aspartame, such as Coca Cola®, and adjust the pH to 5 or 6 with 1 N NaOH using a pH meter. For example, take a 100-ml beaker and add 5 μg of standard 100 μg saccharin to about 4 ml Coca Cola® that had been adjusted to a pH of 5 or 6. Transfer this solution to a 100-ml volumetric flask, dilute to volume with distilled water, and mix well to obtain a 5-ppm solution.

Obtain a cola drink containing saccharin. A typical cola of this type is Like®, which contains about 250 µg/ml saccharin. A better sample would be a diet cola that contains saccharin, but this may be hard to find.

1. Adjust the pH of the cola to 5 or 6 using 1 N NaOH and a pH meter.
2. Transfer 4 ml (equivalent to 1000 µg) to a 200-ml volumetric flask, dilute to 200 ml with distilled water, and mix well. This gives a 5-ppm solution.

Procedure

Secure each of the single standard mixtures of saccharin and proceed as follows, employing the Cary Model 118 with derivative attachment or a spectrophotometer of identical worth. (The instructions given are based on the belief that the Cary Model 118 is being used.)

1. Transverse with an electronic beam and register an absorption spectra. Read the manufacturer's manual and follow the directions given. Begin at 320 nm and scan down to 260 nm.
2. Go back to the beginning wavelength, but use a different part of the chart paper.
3. Place the function switch to "Deriv." Raise the pen away from the paper.
4. Test different settings of the "Abs. Range" switch and the "scan speed" selector and obtain theoretical spectra. Select settings that will give derivative peaks that are approximately 50% full-scale deflection and give the greatest separation of the first and the second derivative peaks of saccharin.
5. Electronically return the wavelength to approximately 10 nm above the beginning wavelength (320 nm) and then go to exactly 320 nm and stop.
6. Touch the pen to the chart paper and obtain a first and second derivative spectra for each of the standard solutions on different parts of the chart paper.

On the chart paper containing one of the standard spectra, procure the first and second derivatives of a blank.

Look critically at the spectra of the standards and pick the wavelengths at which the first and second derivative peaks should be measured. The peak should be measured from the maxima down to the baseline produced with the blank.

Using identical parameters utilized in prior steps but in a different portion of the chart paper, obtain the first and second derivative spectra of the samples and a blank.

Put together two standard curves for saccharin using the 0-, 3-, 6-, and 9-ml standard solutions. One of the standard curves should be for the first derivative peaks, while the other represents the results from the second derivative spectra. Graph paper should be used and plots of concentration vs. peak height should be made.

Utilize the standard curves to find the concentration of saccharin in fortified and nonfortified samples.

Complete the percentage of recovery of saccharin in the spiked samples using the equation below:

$$\% \text{ Recovery of saccharin} = \text{saccharin found (ppm)/saccharin added (ppm)} \times 100$$

Compare the saccharin found and the corresponding percentage of recovery when using the standard curve prepared from the first derivative curves to that found when using the second derivative curves. Compare the precision and accuracy of these types of analysis.

Discussion

This experiment allows the participant to compare analytical results obtained when using first derivative curves to that obtained with second derivative curves. This experiment shows that an inappropriate derivative method can result in error and that each analysis must be judged individually in choosing an appropriate derivative method.

Saccharin has been named orthobenzoic sulphimide by the British Pharmacopeia. Its structural formula is[52]

Saccharin and saccharin sodium are used in the medical profession because their taste is similar to sugar and, therefore, can be used by diabetics, people suffering from obesity, and in other conditions in which sugar should not be used. Sugar is any in a class of sweet, soluble, crystalline carbohydrates such as the disaccharides (sucrose, lactose, and maltose) or monosaccharides (glucose and fructose). Sucrose is the sugar used mainly in the manufacture of food and is prepared from sugarcane and sugar beets. Saccharin is between 200 and 700 times sweeter than the same concentration of sucrose.[53]

When ingested, saccharin is quickly absorbed into the bloodstream and removed from the body via the kidneys, with no changes in its structure. Some rats developed lymphosarcomas when given saccharin at the highest dose level,[54] but the U.S. Congress prevented the Food and Drug Administration (FDA) from banning saccharin, as no definite correlation between saccharin and cancer has been found in humans. Saccharin seems to have a small irritating action and may impair digestion if used at high levels.[54]

Questions

1. *How would you define a first derivative spectrum?*

2. *What is meant by the term "second derivative"?*

3. *Why do you think saccharin tastes sweet?*

4. *Why does sugar taste sweet?*

5. *Is there any common chemical characteristic between saccharin, sugar, and aspartame (the ingredient found in NutraSweet®) that would lead one to believe that certain chemical structures are associated with sweet taste?*

6. *Why has aspartame become more popular in usage than saccharin?*

> 7. *If a 5-µg/ml saccharin standard gives a first de-rivative peak of 25 mm when a 100-ml solution of a cola sample containing saccharin gives a first derivative peak of 28 nm at the same wave-length, what is the concentration of saccharin in the sample if the 100-ml solution of the sam-ple contains 5 ml of the original cola?*

Answers to these questions may be found beginning on page 195.

EXPERIMENT 14
ANALYZING A GAS USING DERIVATIVE SPECTROMETRY

Introduction

Many different methods have been utilized to analyze for ammonia. Some of the methods include titration[55-57] and colorimetric types of analysis.[58-60] The spectrophotometric derivative method is based on the ammonium ion being converted to the ammonia molecule by changing the pH to 11. At 25°C, the nitrogen found in the compound is made up of 98% ammonia and 2% ammonium ion. In a closed system, an equilibrium is set up between ammonia in the liquid state and ammonia in the vapor state. At concentrations that are less than normal, the equilibrium vapor pressure is proportional to the concentration of dissolved ammonia gas in the fluid. The ammonia vapor is analyzed using second derivative spectrometry and the results are mathematically related to the amount of ammonia in the liquid state. Derivative spectroscopy can be used to break two spectral lines of very nearly the same wavelength in separate, constituent elements or parts. In this process, the first and second derivatives of an intensity signal are obtained electronically as a function of time, in which wave-length changes linearly with time. The consequence is a signal very much like the first and second derivative of intensity with respect to wavelength. This signal measures the slope in the intensity distribution. If the energy of radiation at a distinct wavelength is determined after the energy passes through a gas sample, a diminished energy indicates that some of the radiation has been absorbed. A first and second derivative at this same wavelength provides defined details on the gas being measured.

Objectives

The objectives here are to obtain a first and second derivative of a gas and to contrast the methods used to obtain derivatives of a gas in com-parison to that used for liquid samples.

Instruments, Solutions, and Reagents

Procure a Cary Model 118 spectrophotometer or a derivative instrument of equal worth (see Section IV).

Prepare about 100 ml of 10% w/v trichloroacetic acid into a graduate and dilute to 100 ml with deionized water. Mix well.

Obtain high-quality reagent-grade ammonia sulfate from any of several companies (i.e., J.T. Baker, Inc., 222 Red School Lane, Phillipsburg, NJ 08865) and make up ammonium sulfate standard solution at a concentration of about 100 μg/ml using the following procedure:

1. Weigh 0.3897 g ammonium sulfate into a 1-l volumetric flask.
2. Dilute to 1 l with deionized water and mix well.

Secure 50% nitric acid by mixing 50 ml of concentrated nitric acid with 50 ml of deionized water and mix.

Obtain anhydrous potassium carbonate of reagent quality and liquid fertilizer that contains ammonia (available at any hardware store).

Prepare fortified sample solutions by adding 0.05 ml of 100 μg/ml standard ammonium sulfate solution to 0.05 ml sample of liquid fertilizer containing ammonia.

Obtain an absorption cell with a 100-mm light path and with quartz window. The dimensions should be about 22 mm O.D. and about 19 mm I.D.

Procedure

Acquire the standard solution of ammonia and treat using the Cary Model 118 spectrometer (or equivalent) with derivative attachment (described below). Assuming that the Cary 118 is utilized,

1. Clean the absorption cell with 50% HNO_3 solution, rinse the cell with deionized water, and dry.
2. Add 0.05 ml of the 100-μg/ml NH_3 standard, 0.8 g anhydrous potassium carbonate, and 0.5 ml of trichloroacetic acid to the cell and close the cell port.
3. Record an absorption spectrum of the ammonia vapor starting at 300 nm and work down to 200 nm.
4. Go back to 300 nm using a new portion of the chart paper.
5. Put the function switch to "Deriv." and move the pen away from the chart paper.
6. Experiment with several positions of the "Abs. Range" switch and "Scan Speed" selector and actually go though the steps involved in obtaining derivative spectra, but do not mechanically chart. Choose

positions of the abs. switch and scan speed selector that will give 50% full-scale deflection.

7. Bring the wavelength back to approximately 10 or 15 nm above the beginning wavelength and then replace the pen at this wavelength.

8. Put the pen point on the chart paper and obtain a derivative spectra of the standard. Pick an appropriate wavelength in which a prominent peak appears and measure the net peak height of the standard minus a reagent blank.

Using the same parameters, obtain a derivative spectra of the 0.05-ml sample, the 0.05-ml fortified sample, and a 0.05-ml reagent blank. Measure the peak height from the maximum peak down to the reagent blank.

Calculate the percentage of ammonia in the sample and the percentage of recovery of ammonia using the following set of equations:

$$
\begin{bmatrix} \text{Ammonia in} \\ \text{sample } (\mu g/ml) \end{bmatrix} = \begin{bmatrix} \dfrac{\text{net peak height}}{\text{Net peak height}} \\ \text{of standard} \end{bmatrix} \times \begin{bmatrix} \text{Conc.} \\ \text{standard } (\mu g/ml) \end{bmatrix} \times \begin{bmatrix} \text{Dilution} \\ \text{factor} \end{bmatrix} \times \begin{bmatrix} \text{final vol} \\ \text{sample} \\ \hline \text{Final vol} \\ \text{standard} \end{bmatrix}
$$

$$
\begin{bmatrix} \text{Ammonia in spiked} \\ \text{sample } (\mu g/ml) \end{bmatrix} = \begin{bmatrix} \dfrac{\text{net peak height}}{\text{Net peak height}} \\ \text{of standard} \end{bmatrix} \times \begin{bmatrix} \text{Conc.} \\ \text{standard } (\mu g/ml) \end{bmatrix} \times \begin{bmatrix} \text{Final vol} \\ \text{sample} \\ \hline \text{Final vol} \\ \text{standard} \end{bmatrix}
$$

$$
\begin{bmatrix} \% \text{ Recovery ammonia} \\ \text{in spiked sample} \end{bmatrix} = \frac{\text{Ammonia in spiked sample } (\mu g/ml) - \text{Ammonia in sample } (\mu g/ml)}{\text{Ammonia added } (\mu g/ml)} \times 100
$$

Discussion

Derivative spectroscopy can extract more information found within a radiation intensity spectral apportionment than can be obtained through straight spectroscopy procedures because of the derivatives of that distribution.

The experiment just completed demonstrates that derivative spectroscopy can be used to analyze gas as well as liquid solutions.

Ammonia is found in a variety of products including fertilizers and ammonia salts such as $(NH_4)_2SO_4$, NH_4OAc, $(NH_4)_2CO_3$, NH_4OH, NH_4NO_3, etc. Ammonia is also found in surface and saline waters.

The derivative methods of analysis are probably applicable to surface and saline waters and in industrial wastes. Using Nessler tubes,[61] a number of volatile amines cause turbidity with Nessler reagents, but some volatile compounds such as certain ketones, aldehydes, and alcohol may cause off-color on Nesslerization. These types of solutions may be analyzed by derivative spectrometry.

The information presented in this experiment shows that the derivative spectrometric method makes available ammonia quantitation similar to that arrived at using other methods. A qualitative aspect is also introduced because this method involves the absorption of UV radiation of the one and only orderly grouping of absorption bands that are characteristic of ammonia.

Questions

1. *Name five other gases that can be analyzed by derivative spectroscopy.*

2. *Propose a derivative method that can be used to analyze for one of the gases mentioned in Question 1.*

3. *Give a brief explanation of the particle and the wave nature of light.*

4. *What is the relationship between the angstrom, the millimicron and the nanometer?*

5. *Define the electronic ground state and the electronic excited states of atom, ion, or molecule.*

Answers to these questions may be found beginning on page 196.

EXPERIMENT 15
FIRST AND SECOND DERIVATIVE QUANTITATION OF GADOLINIUM

Introduction

Nitrate was found to interfere with gadolinium quantitation using $Gd(NO_3)_3$ solutions. UV absorption spectrometry at 270 to 280 nm using first derivative spectrophotometry was found to overcome this problem. Gadolinium has been analyzed in a variety of ways, including those involving destruction[62,63] or the complexation[64,65] of gadolinium. Among the other methods are the electrochemical methods, using the reaction $Gd^{+3} + 3e^- \rightarrow Gd^0$,[66] and the ion conductivity of 0.98- to 6.3-ppm solutions of dysprosium, europium, gadolinium, and samarium in high purity. Thorium oxide[67] uses a carrier distillation with silver chloride in an inactive gaseous mass comprised of argon-oxygen in a 4:1 ratio as a means of analysis. The rare earths, including gadolinium, can all be quantitated through a mutual action of the rare earths with dyes such as xylenol orange,[68] arsenazo (3-[2-arsono-phenylazo]-4,5-dihydroxy-2,7-naphthalenedisulfonic acid,[69] arsenazo III (2,2',1,8-dihydroxy-3,6-di-sulfo-2,7-naphthylenebisazo) dibenzene arsonic acid,[69,70] and alizarin red.[69] Gadolinium, as well as the other rare earths, will react with these compounds when the concentration of the dyes is <100 μg/ml. All rare earths are pH sensitive and most have a tendency to be interfered with by other elements or compounds. For these reasons, gadolinium must be separated from the other rare earths in order to be quantitated by these dyes.

The absorption spectra of the rare earths have long been used both qualitatively and quantitatively. The absorption spectra of gadolinium and other rare earths have been studied when they were combined with chloride, nitrate, perchlorate, or acetate systems. The slit width of the spectrophotometer used in obtaining the absorbance spectra of gadolinium must be reproducible because the difference in band-pass can have a noticeable effect on the absorptivity.[71] Gadolinium III perchlorate, as well as most of the other gadolinium complexes, absorbs maximally at about 272 nm; however, most rare earths absorb near this wavelength and, therefore, cause interference. X-ray fluorescence has been used for the qualitative detection and semiquantitative analysis of gadolinium in addition to other rare earths.[72] Gadolinium is determined at 2.046 Å.

Atomic absorption spectrometry (AAS) can be used to determine gadolinium and other rare earths.[72] Gadolinium absorbs at 3684.13 Å.

The rare earths form enduring monoxide in air-acetylene or oxyacetylene flames, and therefore can be quantitated by flame photometry.[73] Gadolinium emits at 4058.23 and 4401.85 Å.

Fluorescence has been used to analyze for gadolinium and other rare earths. The rare earths can be excited through electron beams, X-rays, or UV light and fluorescence spectra recorded by a fluorescence spectrophotometer.

Objectives

The objectives of Experiment 15 are to obtain a first and second derivative of a rare earth as well as information on the properties of the rare earths.

Instruments, Solutions, and Reagents

Secure a Cary Model 118 spectrophotometer or another instrument that can obtain derivative spectra. (See Section IV for other derivative spectrophotometers.)

Obtain $Gd(NO_3)_3 \cdot 6H_2O$ from Research Chemicals Canada and prepare $Gd(NO_3)_3$ standard solutions in the following way:

1. Weigh up 100 g of $Gd(NO_3)_3$ into a 1-l flask and dilute to 1000 ml with distilled water. Mix well to obtain a 0.1-g/ml solution.

2. Into separate 1-l flasks, pipette 70, 80, 90, and 100 ml of the 0.1-g/ml solution and dilute to 1 l to get possession of 7-, 8-, 9-, and 10-g/l $Gd(NO_3)_3$ standard solutions. Mix well.

In order to test the method, prepare a standard solution containing 8.5 g $Gd(NO_3)_3$/l by pipetting out 8.5 ml of the 0.1-g/ml solution into a 1-l flask, diluting to 1000 ml with distilled water, and mixing well. This will be the "sample".

Procedure

Obtain derivative spectra of an 8-g/l $Gd(NO_3)_3$ solution using a spectrophotometer with derivative attachment. (The instructions that follow apply to the Cary Model 118 spectrophotometer with derivative attachment. If another instrument is used, follow the manufacturer's instructions.)

1. Register a UV absorption spectra between 350 and 250 nm using the procedures found in the manufacturer's manual.

2. Return the spectrophotometer to 350 nm, but roll the chart paper so that the recordings will be on an unused portion of the chart paper.

3. Raise the pen away from the paper and set the function switch to "Deriv."

4. Test dissimilar settings of the "Abs. Range" switch and "Scan Speed" selector by obtaining spectra between 250 and 350 nm. Choose settings that give peaks at about 50% full-scale deflection.

5. Change the wavelength to about 10 nm above the beginning wavelength and then run the wavelength down to the starting point.

6. Place the pen point on the chart paper and obtain a second derivative spectra. The maximum peak occurs at about 272.8 nm.

Use the same procedure to obtain second derivative spectra for all the standards and the sample using fresh portions of the chart paper each time. Also, obtain the second derivative spectra of distilled water used as the blank.

Prepare a standard curve for $Gd(NO_3)_3$ by locating points on a graph utilizing concentration plotted against the net peak height. The net peak height is the peak height of the standard or sample minus the peak height of the distilled water blank at the same wavelength.

Employ the standard curve to find out the amount of $Gd(NO_3)_3$ in the sample in grams per liter and calculate the percentage of recovery of $Gd(NO_3)_3$ in the sample using the following equation:

$$\% \text{ Recovery of } Gd(NO_3)_3 = \frac{\text{g/l } Gd(NO_3)_3 \text{ found}}{\text{g/l } Gd(NO_3)_3 \text{ added}} \times 100$$

Discussion

This experiment demonstrates that derivative spectrometry can be used for the analysis of rare earths, even in the presence of an interfering object such as nitrate ion.

The rare earths are a number of elements having similar chemical and physical properties. Their atomic numbers are 57 to 71 and constitute only 0.008% by weight of the Earth's crust.[74] The heavy rare earths, gadolinium to lutetium plus yttrium, occur in xenotine, gadolinite, and euxenite commercial ores.

The rare earths are similar in appearance to steel and are soluble in dilute mineral acids. Gadolinium has atomic number 64, atomic weight 157.25, melting point of 1312°C, and boiling point of 3000 K.

Nuclear reactors are customarily provided with two separate emergency shut-off rods or the introduction of a neutron absorber in solutions. Gadolinium nitrate has been used for the latter function.[75]

The characteristic valence of the rare earths, including gadolinium, is 3. The rare earths form complexes with chelating agents such as tartaric acid, citric acid, and ethylene diamine tetraacetate.

Gadolinium and europium are utilized in red phosphors used in color television tubes. Yttrium and gadolinium are used as the host and europium is the activator.

Questions

1. *The rare earths constitute as a group only 0.008% of the weight of the Earth's crust. How much rare earth in grams is found in 1 million lb of that crust?*

2. *Gadolinium has a density of 7.886 g/ml. If a sample of pure gadolinium has a volume of 1.64 cm^3, what is its weight?*

3. *Some pure metal samples contained in beakers lost their labels. One sample was found to have a volume of 2 ml and weighed 14 g. The metal was thought to be gadolinium, europium, samarium, or promethium (densities of 7.886, 5.245, 7.536, and 7.016 g/cm^3, respectively). What was the identity of the metal?*

4. *Describe the atomic structure of gadolinium, including its outer electron configuration, and explain how gadolinium undergoes chemical reactions.*

5. *Gadolinium has the following properties: atomic number, 64; atomic weight, 157.25; probable outer electronic configuration, $4f^7\ 5d^1\ 6s^2$; density, 7.886 g/ml; atomic volume, 19.88 ml/mol; melting point, 1312°C; and boiling point, 300°C. Name a method that could be used to separate gadolinium from a mixture of samarium and europium.*

6. *Gadolinium nitrate absorbs maximally at about 272 nm in a broad peak. Draw a hypothetical absorption spectra for this compound between 350 and 250 nm.*

Answers to these questions may be found beginning on page 198.

EXPERIMENT 16
ANALYSIS OF DRUGS IN THE PRESENCE OF THEIR DECOMPOSITION PRODUCTS. THE CASE OF PHENACETIN

Introduction

Acetaminophen (*n*-acetyl-*p*-aminophenol) can break down into *p*-aminophenol and acetic acid[76] in the following way:

$$HO{-}\bigcirc{-}NHCCH_3 \quad \xrightarrow[\text{H}^+\text{ or (OH)}^-]{\text{H}_2\text{O}} \quad CH_3COH \; + \; HO{-}\bigcirc{-}NH_2$$

The reaction can be implemented by acids and bases,[77] which can both act as a catalyst. As a contrast to this, phenacetin (also called acetophenetidin) is hydrolyzed by acids to form *p*-phenetidine and acetic acid.[78]

Diverse methods have been used to determine *p*-aminophenol as a breakdown product of acetaminophen. Spectrophotometric methods have been founded on the formation of colored solutions of *p*-aminophenol with certain reagents. Aminophenol is usually separated prior to analysis by one of the chromatographic methods such as column chromatography or TLC. Aminophenol has also been determined by fluorometry. Aminophenol contains the electron donating groups $-OH$ and $-NH_2$, both of which increase the fluorescence. Other systems include liquid chromatography, polarography, and nonaqueous titration with perchloric acid.[79]

Some analyses determine the total drug content of phenacetin (acetaminophen) and its breakdown products. This can be achieved by using phenol to form indophenol. The indophenol is then reacted with other reagents to form a colored compound that is then determined spectrophotometrically.[80]

Prior to the present time, derivative spectroscopy has been put to some practical and specific use in the analysis of several drugs in the vicinity of their degradation products. One analysis involves ascertaining procaine in the presence of 4-amino benzoic acid.[81] 1,4-Benzodiazepines have been analyzed in the immediate vicinity of their degradation products. The degradation products of this compounds are acid catalyzed.[82] Thiamin and pyridoxamine were analyzed in close proximity of the degradation products that formed as the compounds aged.[83] Cephalosporins have been quantitated without separation of their decomposition products.[84]

Derivative spectrometry has been utilized for the analysis of decomposition products immediately surrounded by nondegraded drugs. An example is the analysis of salicylic acid in the presence of aspirin[85–87] and the determination of sulfoxide as a breakdown product of chlorpromazine.[88]

Objectives

The objectives here are to illustrate how derivative spectrometry can be used to analyze for an intact drug, phenacetin, in the presence of its decomposition products and to acquire knowledge about the chemical and physical characteristics of acetaminophen.

Instruments, Solutions, and Reagents

Obtain a derivative spectrophotometer such as the Cary Model 118. (See Section IV for other derivative spectrophotometers.)

Obtain phenacetin from the USP.

Prepare *p*-phenetidin standard solution by reacting phenacetin with HCl in the following way:

1. Weigh about 100 mg phenacetin standard into a small flask and add 10 ml of concentrated HCl. Attach a water condenser and reflux about 30 min at 100°C over boiling water. Cool and transfer the solutions to a 100-ml volumetric flask with 10 ml of water and then dilute to 100 ml with 0.1 *N* HCl to obtain a 1000-μg/ml solution.

2. Take a 1-ml aliquot (equivalent to about 1000 μg) into a 100-ml volumetric flask and dilute to 100 ml with 0.1 *N* HCl to obtain a 10-μg/ml solution.

Prepare phenacetin standard solutions in the following way:

1. Weigh about 100 mg of phenacetin into a 100-ml volumetric flask and add 5.0 ml of methanol to dissolve. Dilute to 100 ml with 0.1 *N* HCl to obtain a 1-mg/ml solution.

2. Into eight separate 100-ml volumetric flasks, pipette 0, 0.5, 1.5, 2.0, 2.5, 3.0, 3.5, 4.0, and 5.0 ml of the 1-mg/ml solution. Dilute to 100 ml with 0.1 *N* HCl to obtain 0.5, 1.5, 2.0, 2.5, 3.0, 3.5, 4.0, and 5.0 mg/100 ml.

For the purpose of testing the derivative method, obtain acetophenetidin tablets.

1. Obtain the average weight of 20 tablets, crush, and pass through a sieve to get rid of large particles.

2. Weigh out two samples of the crushed tablets, each equivalent to 100 mg of phenacetin, into two separate 100-ml volumetric flasks. Add 5 ml of methanol to dissolve. Dilute to 100 ml with 0.1 N HCl to obtain a 1-mg/ml solution.

3. Into two separate 100-ml flasks (one for each sample), pipette 3.0 ml of approximately 1-mg/ml solution and dilute to 100 ml with 0.1 N HCl.

Procedure

Procure the 30 mg/ml phenacetin solution and obtain spectra using a spectrophotometer with a derivative attachment. (The information given below assumes the use of the Cary Model 118 spectrophotometer with derivative attachment.)

1. Record a UV absorption spectra in the region of 250 and 325 nm using the steps found in the manual provided by the manufacturer.

2. Bring the wavelength back to 325 nm but change the chart paper so that the pen will be set on a different part of the chart paper.

3. Lift the recorder pen from the chart paper and set the function switch to "Deriv."

4. Examine different adjustments of the "Abs. Range" switch and "Scan Speed" selector by securing spectra in the region of 250 to 325 nm. A wavelength of about 260 nm should prove to give acceptable spectra with peaks at about 50% full-scale deflection.

5. Scan up to about 5 or 10 nm above 325 nm and then back to 325 nm.

6. Place the recorder pen on the chart paper and run a derivative scan between 325 and 250 nm.

Utilize the same method to obtain a second derivative spectra for all the standards and the samples utilizing a different portion of the chart paper.

Prepare a blank solution by placing 5 ml of methanol into a 100-ml volumetric flask and diluting to volume with 0.1 N HCl. Obtain a second derivative spectra of the blank by scanning between 325 and 250 nm.

Create a calibration curve for phenacetin by plotting concentration vs. the net peak height of the derivative peak. The net peak height can be obtained by subtracting the peak height of the sample from the peak height of the blank at the same wavelength. Using the calibration curve, determine the concentration of phenacetin in the sample.

Prepare a degradation solution of phenacetin by weighing about 100 mg of phenacetin into a 50-ml volumetric flask:

1. Dissolve in 6 N HCl and dilute to volume with the same solvent.

2. Mix contents and place in a constant temperature water bath at the same temperature at which the phenacetin samples were analyzed.

3. After about 5 min, let the solution equilibrate thermally and transfer 2 ml, representing the sample at zero hour, to a 100-ml volumetric flask. Cool to room temperature and dilute to volume with distilled water.

4. Similarly, transfer a 2-ml aliquot into another 100-ml volumetric flask and dilute to 100 ml after 15 min.

5. Obtain the second derivative spectra between 325 and 250 nm.

6. Calculate the concentration of phenacetin in each of the degradation samples using the standard curve.

Discussion

The degradation product of phenacetin does not give any appreciable interfering peaks in the second derivative spectrum. This shows that phenacetin can be quantitated even in the presence of comparatively large amounts of the degradation products.

Another name for phenacetin is acetophenetidin. There are several ways to make phenacetin. In one method, phenol is changed to *o*- and *p*-nitrophenol. The *ortho* derivative is removed by steam distillation. *p*-Nitrophenol is converted to the sodium derivative and then reacted with ethyl chloride or ethyl bromide to create ethyl *p*-nitrophenol. Reduction is then obtained with sodium or iron fillings along with HCl to form *p*-phenetidin. Glacial acetic acid is then used to introduce the acetyl group to produce *p*-acetophenetidin.

Phenacetin occurs as white crystals;[89] it is odorless, slightly bitter, stable in air, and relatively insoluble in water but quite soluble in alcohol and chloroform. Phenacetin is utilized for its analgesic and its antipyretic properties. It is often found in combination with other drugs such as acetanilid, caffeine, codeine sulfate, etc. Both acetanilid and phenacetin will break down in the body in order to produce *n*-acetyl-*p*-aminophenol. Phenacetin and acetanilid have very similar properties.

Analgesics are compounds that rid the body of pain by raising the beginning point of the pain. This beginning point seems to vary with age, sex, circulatory differences, skin temperature, and certain emotions. The action of analgesics even varies in the same individual and is certainly different between individuals.

Phenacetin (acetophenetidin) has the molecular formula $C_2H_5OC_6H_4NHCOCH_3$. Besides its analgesic property, it is also an antipyretic (a compound that reduces body temperature).

Questions

1. *Phenacetin can be analyzed by a variety of methods. Describe the official method found in the USP.*

2. *Phenacetin contains the CH_3 group. Review an organic chemistry book and describe the type of bonds found with this type of carbon.*

3. *Assume that you had some old phenacetin tablets that had obviously broken down. The tablets declare 250 mg per tablet. Give the general steps that you would follow in order to analyze these tablets.*

4. *Phenacetin breaks down into p-phenetidin and acetic acid under the influence of strong acids or bases. Write the equation (using condensed formulas) for this reaction.*

Answers to these questions may be found beginning on page 199.

Answers to these questions may be found beginning on page 199.

EXPERIMENT 17
STABILITY STUDIES OF VITAMINS USING DERIVATIVE SPECTROMETRY

Introduction

Because vitamins tend to lose their potency with age, it is important to use a reliable method to determine this loss with respect to time. Many of the vitamins contain pyridoxine and thiamin hydrochloride in a solid form such as capsules and tablets. Excipients are also a normal part of the mixture. It is perfectly legal for the vitamins to contain excipients such as bases, carriers, coatings, colors, flavors, preservatives, stabilizers, and vehicles. These excipients increase the stability and usefulness of the various products. Most of the present-day methods are not specific enough and do not take into account the effect of interfering substances (such as the excipients) on the result of the analysis.

The USP method for vitamins uses vitamins that are free of interferences as models. The same can be said for the BP.[90,91] Obtaining thiamin in a pure state involves protracted and irksome methods such as column, paper, or thin layer chromatography.[92,93] In the USP, pyridoxine is analyzed by reacting it with 2,6-dichloroquinone-chlorimide[94] to form a colored compound that is then run on a spectrophotometer. The problem is that the color is not specific for pyridoxine because other compounds will give the same color, and in many cases the color will lessen.

Both pyridoxine and thiamin have distinctive UV absorption spectra, but because of interference from other vitamins in multivitamin preparations or from excipients, UV absorption has not been generally used for the analysis of these compounds. Another source of interference is a result of the breakdown of the vitamins themselves. Derivative spectrometric measurement of pyridoxine and thiamin preparations over a period of time has the advantage of intensifying the specificity and precision of the analysis by recording the peaks at selected wavelengths with the spectral interferences eliminated.

Objectives

The objectives of this experiment are to demonstrate how derivative spectrometry can be used to determine in what ways the potency of vitamins changes with time, and to become more knowledgeable about vitamins.

Instruments, Solutions, and Reagents

Procure a derivative spectrophotometer such as the Cary Model 118. (See Section IV for other derivative spectrophotometers.)

Acquire pyridoxine and thiamin hydrochloride standards from the USP.

Obtain samples of thiamin and pyridoxine hydrochloride dating back several years. (One way would be to write a prominent drug manufacturer such as Parke-Davis and request that they send these vitamins to aid in your research project.)

Acquire reagent-grade dimethyl ether, chloroform, and phosphoric acid. Prepare phosphate buffer, pH 7, in the following way:[95] dissolve 15.22

g $KHPO_4$ $3H_2O$ and 4.54 g KH_2PO_4 in water and dilute to 1 1. Check the pH using a pH meter.

Prepare thiamin hydrochloride standard solution at a concentration of 20 µg/ml by complying with these directions:

1. Weigh 20 mg of the standard into a 10-ml volumetric flask and dilute to volume with water and mix well. This solution has a concentration of 2000-µg/ml.

2. Pipette 1 ml of the 2000-µg/ml solution into a 100-ml volumetric flask and dilute to 100 ml with phosphate buffer, pH 7. Mix well. This gives a 20-µg/ml standard solution.

Prepare pyridoxine hydrochloride standard solution at a concentration of 10 µg/ml using the procedure described below:

1. Weigh 10 mg of the standard into a 10-ml volumetric flask and dilute to volume with deionized water and mix well. This solution has a concentration of 1000 µg/ml.

2. Pipette 1 ml of this solution into a 100-ml volumetric flask and dilute to volume with phosphate buffer, pH 7, and mix well. Thus, a 10-µg/ml solution is obtained.

Assuming that tablets manufactured in previous years have been obtained, separate the average weight of 60 tablets from each of these years. Grind each group of 60 tablets individually and put them through a #60 sieve.

Prepare two or more thiamin hydrochloride samples using the following procedure:

1. Weigh out the equivalent of 200,000 µg thiamin hydrochloride into a 100-ml centrifuge tube containing 40 ml of chloroform and shake for 5 min.

2. Add 20 ml of dimethyl ether and centrifuge for 20 min.

3. Use a pipette to remove the top layer in which the desired vitamin is found and place the supernatant into an appropriate size beaker.

4. Add another 40 ml of chloroform and shake 5 min. Add 20 ml of dimethyl ether, centrifuge, and remove the supernatant as before.

5. Evaporate the extracted solution to dryness in a steam bath.

6. Add 40 ml of hot water and boil for 30 min while maintaining the volume.

7. Centrifuge the aqueous solution and transfer this solution to a dark-colored 100-ml volumetric flask.

8. Re-extract the remaining residue with 40 and 20 ml of boiling water as done in part 6 above.

9. Pipette 1 ml of this 2000-µg/ml thiamin hydrochloride sample solution into another 100-ml volumetric flask and dilute to volume with phosphate buffer solution, pH 7, to obtain a 20-µg/ml solution.

Prepare two or more pyridoxine hydrochloride samples in the same way, except the initial concentration in the first 100-ml volumetric flask should be 1000 μg/ml (refer to part 1 above) and the final concentration in the last 100-ml volumetric flask should be 10 μg/ml (see part 9).

Procedure

Acquire a derivative spectrophotometer. There are many manufacturers of derivative spectrophotometers, but the directions that follow specify use of the Cary Model 118 spectrophotometer with derivative attachment. Thiamin hydrochloride at 20 μg/ml is utilized first, but the directions also apply to pyridoxine hydrochloride at 10 μg/ml.

1. Obtain a UV absorption spectrum between 200 to 375 nm utilizing the manufacturer's directions.

2. Use a fresh portion of the chart paper and scan back to 375 nm.

3. Set the derivative switch to "On" but raise the pen away from the chart paper.

4. Obtain imaginary spectra using various settings of the "Scan Speed" selector and the "Abs. Range." Look for wavelengths between 200 and 375 nm that give good spectra at about 50% full-scale deflection.

5. Bring the pen back to about 5 or 10 nm above 375 nm and then down to 375 nm.

6. Set the pen on the chart paper and run a first and then a second derivative scan.

Use this procedure to obtain second derivative spectra of the samples and standards.

Make up a blank solution using 1 ml of water diluted to 100 ml with distilled water. Obtain first and second derivative spectra using the same procedure that was used with the samples and standards.

Measure the appropriate peak height of the sample and standards by measuring from the peak down to the blank baseline.

Calculate the concentration of thiamin hydrochloride and pyridoxine hydrochloride using the formula:

$$\begin{pmatrix} \text{Conc. vitamin} \\ \text{in sample} \end{pmatrix} = \frac{[\text{Net peak height of sample}]}{[\text{Net peak height of standard}]} \times \begin{bmatrix} \text{Conc. std.} \\ (\mu\text{g/ml}) \end{bmatrix} \times \frac{\begin{bmatrix} \text{Final vol} \\ \text{of Sample} \end{bmatrix}}{\begin{bmatrix} \text{Final vol} \\ \text{of standard} \end{bmatrix}} \times \begin{bmatrix} 1 \text{ Wt.} \\ \text{Sample (g)} \end{bmatrix}$$

Discussion

Thiamin hydrochloride and pyridoxine hydrochloride have UV absorption spectra that give peaks at 268 and 325 nm, respectively. This means that they can be analyzed in the presence of each other using this method if allowances for minor interferences are made. However, derivative spectrometry is an alternative method that has many advantages. One advantage that derivative spectrometry has over UV absorption spectrometry is that the UV absorption of thiamin and pyridoxine severely overlap below 260 nm and cannot be analyzed by this procedure. However, derivative spectrometry can be used.

Nutrition is the science that identifies substances called nutrients that are needed for health. Nutritionists determine the amounts of these nutrients that are required. For the last 30 years, the Food and Nutrition Board of the National Academy of Science has published recommended dietary allowances (RDA). The RDA are derived from a variety of combinations of foods. No one food contains all of the desired nutrients. There is a strong health risk when an individual goes on a single food diet or single type of food diet. For example, some diets consist of eating watermelon or drinking milk as the only source of food intake. The best diet is varied, and includes dairy products, proteins, vegetables, and carbohydrates.

Nutrients that are required by the human adult include: oxygen, water, energy from glucose or fat, protein in the form of the essential amino acids (tryptophan, phenylalanine, lysine, threonine, methionine, leucine, isoleucine, and valine), sodium chloride, potassium, phosphorus, calcium, magnesium, and the essential fatty acids.[96]

Other nutrients required by humans include iron, vitamin C, zinc, nicotinic acid amide, vitamin B_3, fluorine, vitamin E, copper, vitamin B_5, manganese, vitamin K, vitamin B_6, vitamin A, vitamin B_2, vitamin B_1 (thiamin), iodine, folic acid, biotin, vitamin D, chromium, vitamin B_{12}, molybdenum, vanadium, selenium, and tin.

There are two classes of vitamins, fat soluble and water soluble, and both are organic compounds. One, however, is nonpolar (fat soluble) and the other is polar.

The fat-soluble vitamins include vitamins A, D, E, and K. These vitamins tend to accumulate in the fatty tissue of our bodies. Vitamin A has the following structure:

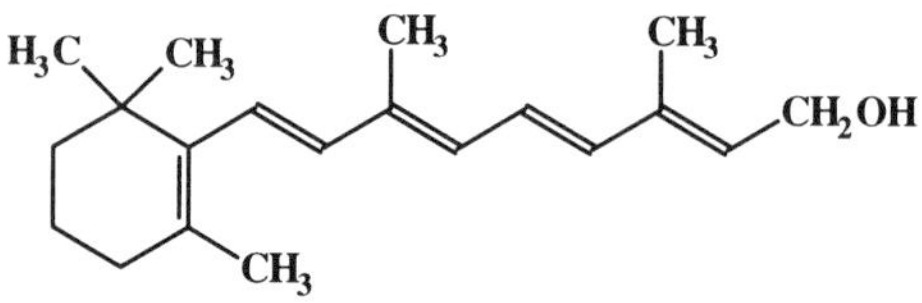

Vitamin A is needed for healthy mucous membranes and healthy eyes. Extensive deficiency of vitamin A can lead to blindness. Excessive amounts of vitamin A can lead to illness.

Vitamin D aids in the formation of bones and teeth. Lack of vitamin D can cause rickets. Sunlight converts the steroids in our skin into vitamin D. Vitamin D has the following structures:

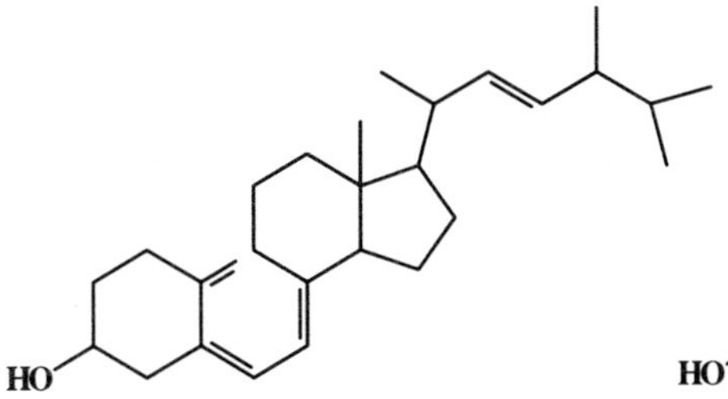

Ergocalciferol (D$_2$) 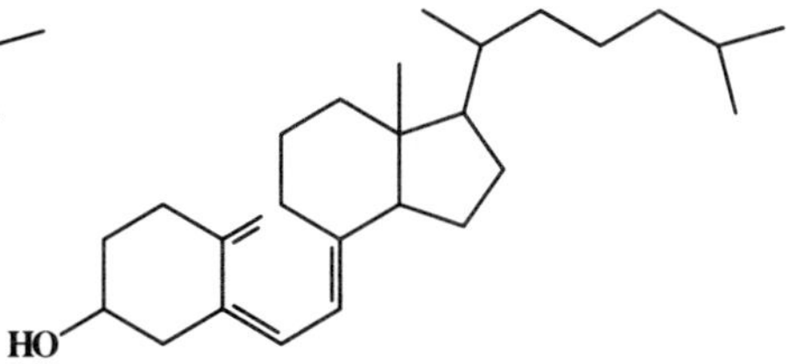   **Cholecalciferol (D$_3$)**

Vitamin E is found in vegetable oils. Americans do not seem to be deficient in this vitamin. Vitamin E has this construction:

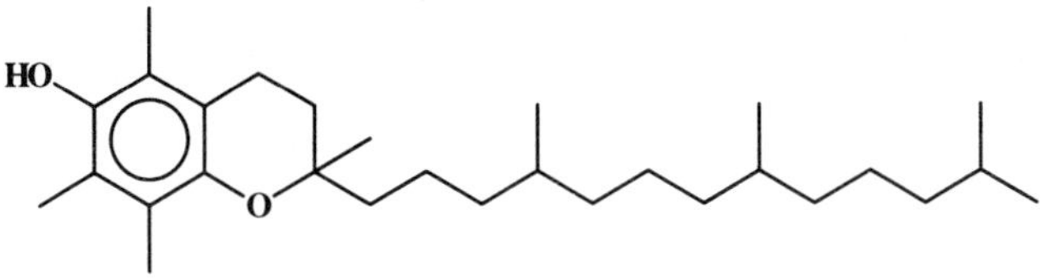

The structure of vitamin K is

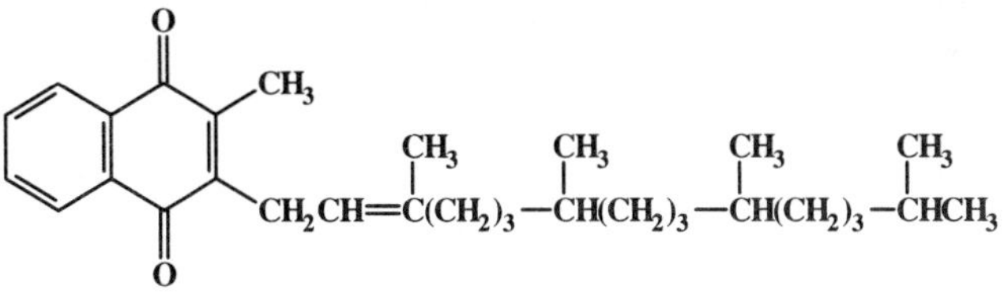

Vitamin K is found in leafy vegetables. It is an antihemorrhagic vitamin which the American diet does not seem to lack.

The water-soluble vitamins include vitamin C, choline, thiamin, riboflavin, niacin, vitamin B$_6$, vitamin B$_{12}$, pantothenic acid, and biotin.

Vitamin C acts to prevent scurvy, a disease in which collagen is poorly manufactured. In addition, vitamin C plays a part in the body's manufacture of some adrenal hormones, the healing of wounds, and the metabolism of amino acids. Vitamin C is found in citrus fruits, potatoes, and leafy vegetables. Vitamin C has the following structure:

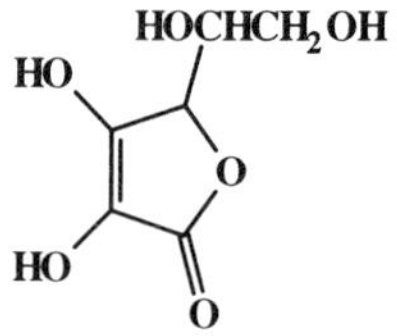

The function of choline is to manufacture certain lipids. Acetylcholine helps to transmit nerve signals. It can be found in meats, eggs, cereals, and beans. Its structure is as follows:

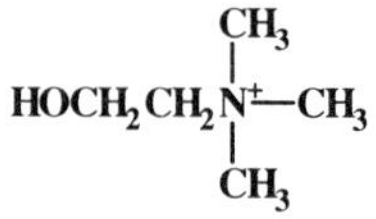

Thiamin is found in meats, beans, and whole grains. There are two types of vitamin B_1 deficiency. One is known as dry beriberi, in which the predominant symptoms involve the nervous system, especially the peripheral nerve fibers of the lower extremities. Degenerative changes are found in the ganglion cells of the nervous system with degeneration of the myelin sheath and fragmentation of the axis cylinder of the peripheral nerves and atrophy and degeneration of the muscles supplied by these nerves. Malfunction of the circulatory system characterizes the second form of beriberi. Thiamin has the following structure:

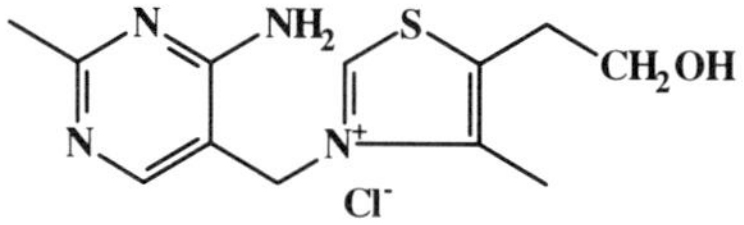

Riboflavin is principally found in milk. It has the structure

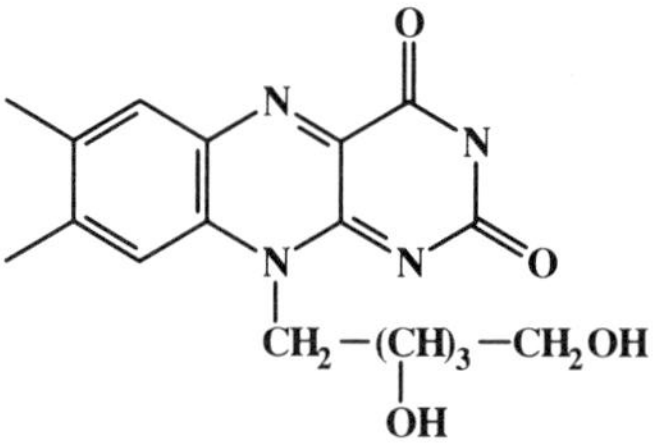

Riboflavin is necessary for almost all biological oxidations. A deficiency of riboflavin affects wound healing and leads to breakdown of the skin around the mouth, nose, and the tongue.

Niacin encompasses both nicotinic acid and nicotinamide. This vitamin is involved in most biological oxidations. A lack of the vitamin causes pellagra, in which there is a breakdown of the nervous system and various skin problems. It can be found in meat. The structures are

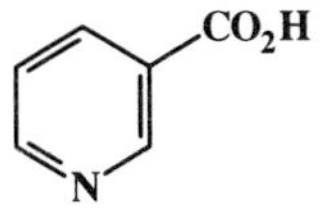

Nicotinic Acid 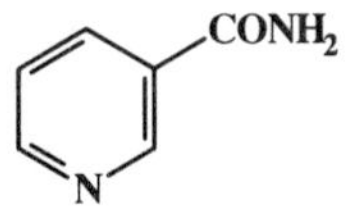   **Nicotinamide**

Folic acid or folacin is a vitamin used for the synthesis of nucleic acid and heme. Lack of sufficient amounts of this vitamin causes megaloblastic anemia. Natural sources of this vitamin can be found in leafy green vegetables, asparagus, liver, and kidneys. Its structure is

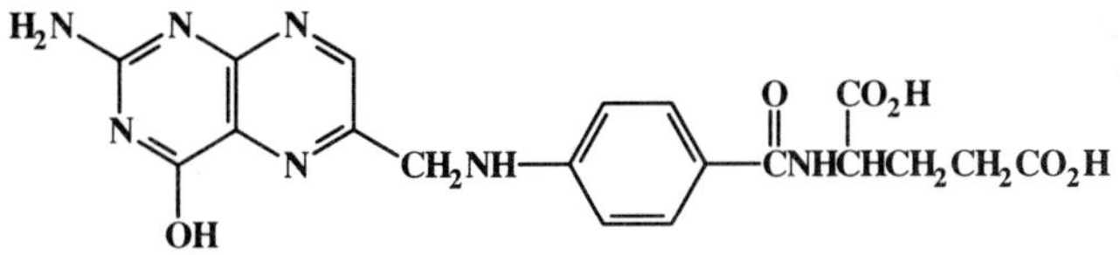

Vitamin B_6 is pyridoxal phosphate. It is used primarily in the metabolism of amino acids and can be found in meats, yeasts, and corn. When lacking in the body, a certain type of anemia develops as do malfunctions of the nervous system. The structure of vitamin B_6 is given below:

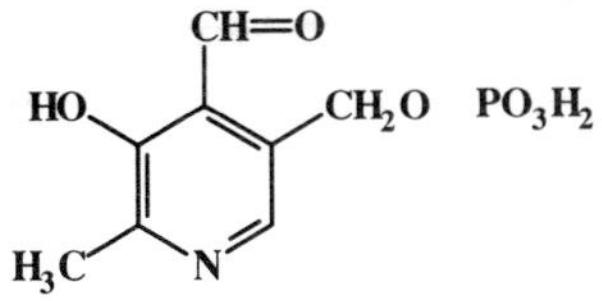

Vitamin B_{12} is one of the few vitamins that strict vegetarians *(vegans)* lack in their diet. It is found in lean meats and milk products as well as in eggs. Deficiencies of this vitamin cause pernicious anemia. Its structure is given below:

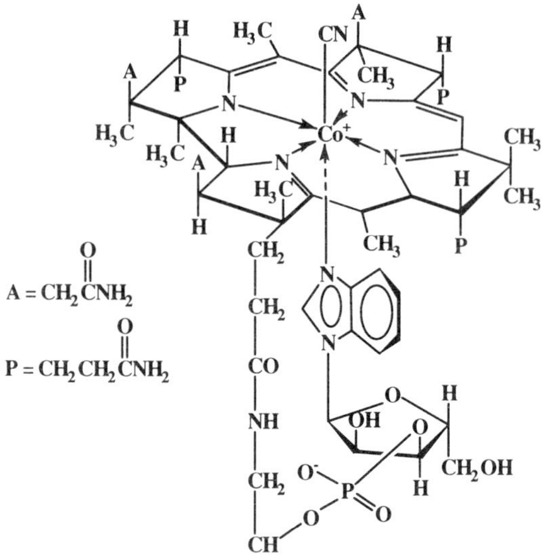

$$A = CH_2\overset{\overset{\displaystyle O}{\|}}{C}NH_2$$

$$P = CH_2CH_2\overset{\overset{\displaystyle O}{\|}}{C}NH_2$$

Pantothenic acid produces an enzyme that is used to metabolize fatty acids. Lack of pantothenic acid causes cell damage. The following compound is pantothenic acid:

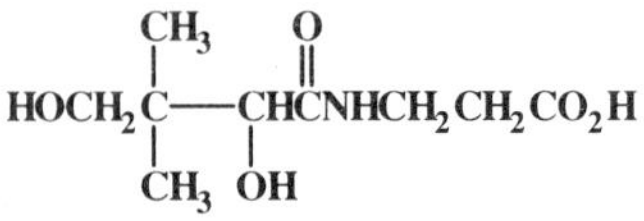

Biotin deficiency causes anorexia, nausea, pallor, dermatitis, and depression. Biotin is required in cases in which carbon dioxide is used, for example, in the synthesis of fatty acids by the body. Biotin has the structure listed below:

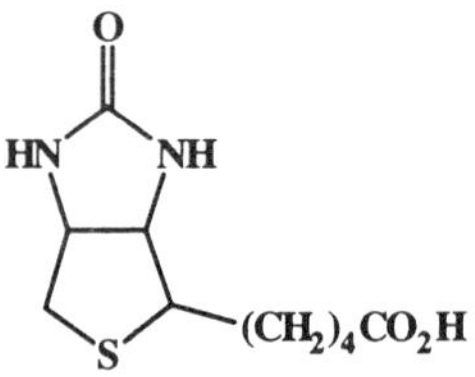

Questions

1. *Tell how derivative spectrometry can be used to analyze a mixture of vitamins D and E.*

2. *Assume you have a solution of a vitamin that you suspect is pyridoxine hydrochloride. Tell how you could use derivative spectrometry to qualitatively identify this compound.*

3. *Why should our daily diet be drawn from a variety of foods?*

4. *On a strict vegetarian diet, what one vitamin is hardest to obtain?*

5. *Derivative spectrometry is based on absorption of energy. Explain the absorption process in terms of the wave nature of light.*

Answers to these questions may be found beginning on page 200.

Section II

Application of Derivative Techniques to Pesticide Analysis

INTRODUCTION

The chemicals called pesticides are categorized according to their function. These functions include the control of insects (insecticides), weeds (herbicides), and fungi (fungicides). Other pesticides are called fumigants, larvacides, miticides, insect repellents, rodenticides, plant-growth regulators, and defoliants. Pesticides have several beneficial characteristics, including increasing the world's food supply and controlling diseases, as many diseases are transmitted by pests. Some authorities claim that pests are responsible for about half the human deaths because of the diseases they carry and the destruction or consumption of one third of the human food supply.

Insecticides include the chlorinated variety, such as aldrin, heptachlor epoxide, dieldrin, endrin, and p,p' DDT. These insecticides are characterized as being very resistant to decomposition into less toxic compounds. Thus, they pose a hazard to animals, birds, fish, and mammals, including humans.

Organophosphates are another important class of insecticides. Some common organophosphates are methamidiphos, diazinon, chlorpyrifos, acephate, malathion, methyl parathion, and azodrin. These compounds are more susceptible to environmental forces and tend to break down due to hydrolysis and oxidation. The organophosphates act by blocking the active site of the enzyme acetylcholinesterase. This enzyme is found in both insects and humans. Even though these insecticides are not as long lived as the organochlorine compounds, they still pose a health hazard to many forms of life.

The carbamate insecticides are represented by carbaryl. Some important herbicides include 2,4-dichlorphenoxy acetic acid, atrazine, and simazine. An important example of the fungicides is the so-called Bordeaux mixture, which is made from lime and copper sulfate, organics of the thiocarbamate class, and mercury and tin compounds.

The authority to regulate and monitor pesticides lies with three agencies of the federal government. One, the Environmental Protection Agency (EPA), registers and approves a given pesticide and sets tolerances. The Food and Drug Administration (FDA) enforces the tolerances set by the EPA and obtains data on the range of occurrence and quantities of pesticide residue in all goods or commodities subject to interstate transport or that are presented for entrance into the U.S., except those that fall under the responsibility of the U.S. Department of Agriculture (USDA). The USDA is accountable for regulating pesticides in some egg products, poultry, and meat.

At the present time, most agencies concerned with pesticide monitoring are looking for better testing methods for pesticide residue. New approaches to pesticide residue testing will probably supersede the methods presently accepted, which involve extraction and gas chromatography. It is probable that some type of analysis will be done in the field and only those samples that test positive will be further tested.

Derivative techniques may become one of the new methods applied to the analysis of pesticides. To illustrate, some examples are offered of how first and second derivatives can be used to quantitate peaks produced by liquid chromatographic (LC) methods using various types of detectors, including UV and fluorescence.

The first example (see figure below) shows two symmetric peaks that are well separated. The first and second derivatives are also shown. A well-separated peak has a singular appearance in the second derivative. It rises to maximum, falls to minimum, rises to a second maximum, and returns to zero. This singular appearance may be distorted by tailing or another nonsymmetry, but the above named three constituents will still be present in the order given. If this sequence is not followed, this indicates that this is not a normal peak. Each of the maximum peaks of the original spectra has a minima associated with it in the second derivative spectra.

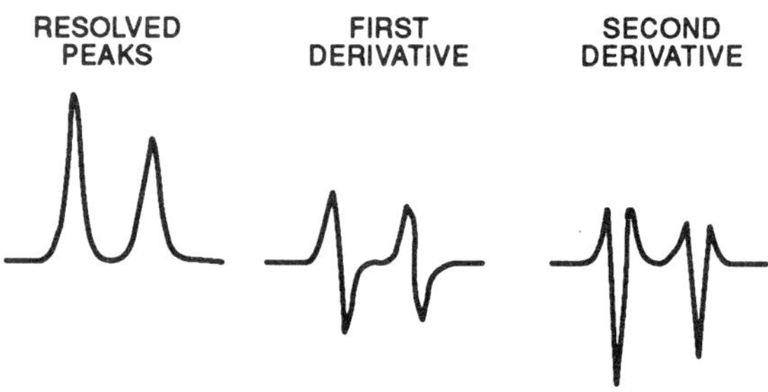

In the next figure, peaks A and B are closer together. On examining the second derivative, the characteristic peaks observed with the well-separated peaks are no longer present. For peak A, the second derivative rises to maximum, falls to minimum, rises to a second maximum, but does not return to zero. This indicates the presence of either a valley point or a shoulder. The appearance of the first derivative indicates that a valley exists. In cases in which the LC peaks have begun to come closer together, a prominent valley occurs between the two peaks. At this valley point, the slope of the peak will change from negative (going down) to positive (going up). The first derivative passes through zero and the appearance of the first derivative at this zero crossing point is characteristic of a valley point. An arrow shows the area of the first derivative, which indicates the presence of a valley in the original peaks.

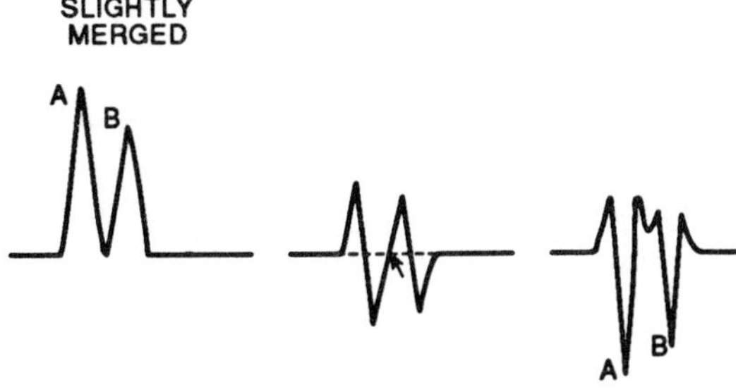

In the next figure, A and B are much closer together. This causes the shapes of the first and second derivation to change drastically. The second derivative minima for peaks A and B have combined into a single maximum, but the first minima for peaks A and B are still very conspicuous. The appearance of the first derivative at the zero crossing point is typical of a valley point and not a shoulder. The arrow indicates the point in the first derivative (the actual crossing of the zero line) that indicates the existence of a valley between the two LC peaks we labeled A and B.

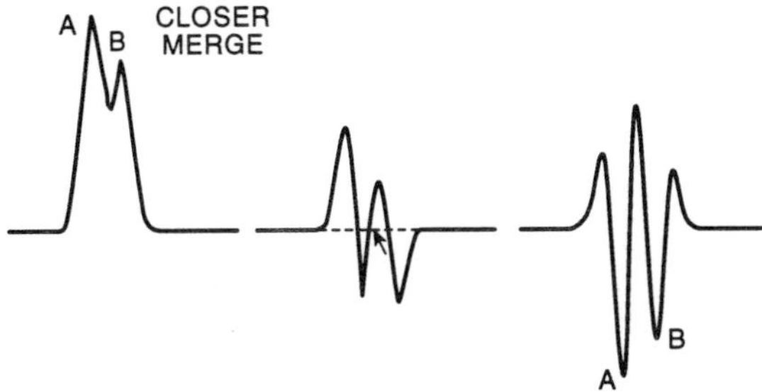

In the next figure, the two chromatographic peaks have come together so that peak B forms a shoulder on peak A. The two large minima seen in the second derivative spectra indicate that there were two chromato-graphic peaks in the original spectrum. The first derivative spectra shows that a shoulder is present, because the first derivative peak does not cross zero following the first minima.

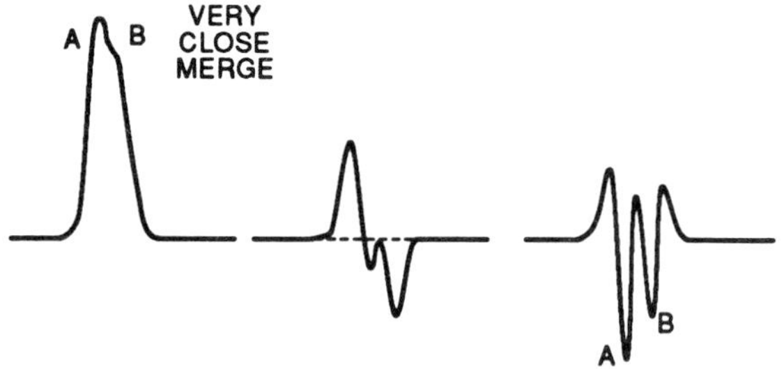

In the next figure, there is a very small shoulder as opposed to the distinctive shoulder seen in the previous example. The second derivative spectra shows a large minima and a small minima representing liquid chromatographic peaks A and B, respectively. The first derivative peak indicates that a shoulder is present in the original chromatographic peaks because the first derivative does not cross the zero line after the first minima.

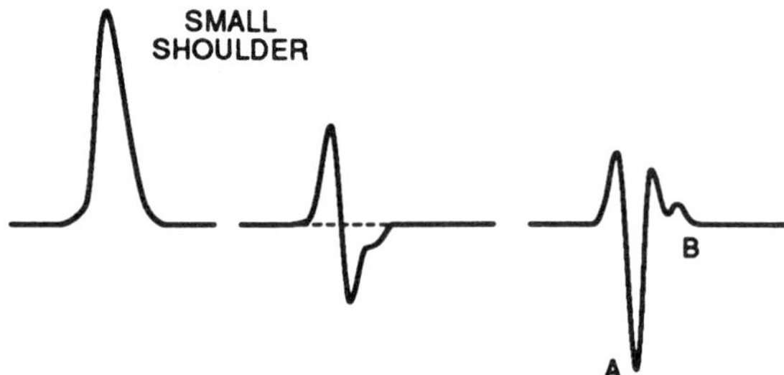

In the last figure, the two peaks are completely combined and the resulting peak is almost indistinguishable from a single peak, except that the second derivative spectra shows two minima that correspond to LC peaks A and B.

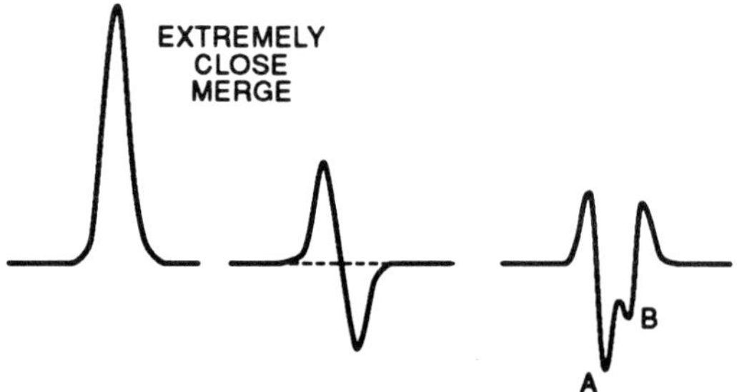

Most of the peak types found in LC in which derivative analysis would be helpful are illustrated by the second figure, in which the peaks are slightly merged, and by the third figure, in which we have a closer merge. It would be much easier to measure the second derivative peaks for quantitative analysis rather than to try to measure the original spectra.

A review of five papers follows in which LC was used to analyze various pesticides. Spectra will be shown in which interfering peaks are present along with the desired analyte. Each section begins with the names of the authors, followed by an abstract or summary of the analysis. Then, a figure is presented in which the analyte of interest is surrounded by interfering peaks, and the problems caused by these interferences are discussed. Finally, a figure is presented showing how the second derivative of the desired analyte might look and a discussion of how it would be easier to quantitate this pesticide by means of a second derivative analysis in comparison to the original method follows.

LC WITH PHOTOCONDUCTIVITY/UV DETECTION APPLIED TO THE DETERMINATION OF CAPTAN
A REVIEW OF AN ARTICLE BY STEPHEN M. WALTERS[97]

Introduction

The applicability of the Tracor Model 965 photoconductivity detector (PCD) to the determination of a variety of pesticide chemicals, particularly polar and/or thermally labile compounds which are troublesome in gas chromatographic (GC) analysis, has been investigated. The effects of various operating parameters (e.g., mobile phase composition, flow rate, and irradiation wavelength) on signal-to-noise output for selected compounds have been evaluated. A comparison of photoconductivity responses with those obtained from a UV detector connected in tandem was made for selected reference standards and for food sample extracts. The photoconductivity detector was found to be suitable for the determination of pesticide residues at subparts per million levels. The linearity and reproducibility of response are adequate for practical quantitative applications.

Analysis by Derivative Spectrometry

The PCD-UV tandem detector system was applied to food sample extracts. A strawberry sample extract was prepared by a method found in the *Pesticide Analytical Manual* (Vol. 1, 2nd ed., Food and Drug Administration, Washington, D.C., 1968, rev. 1982, Section 212.13c). The strawberry extract was analyzed for captan by the UV and the PCD. The PCD gave adequate peaks and the sample was found to contain 0.03 ppm captan. However, the UV peak was so small that no analysis could be made by UV. This peak (shown in Figure 12) could be made less broad and more narrow by adjusting the LC parameters. The second derivative measures the curvature of the signal. Anything that can be done chromatographically, e.g., narrowing very broad peaks, to increase the curvature will increase the ease of detection. Assuming that this could be accomplished, the second derivative of the peak representing captan might look like this:

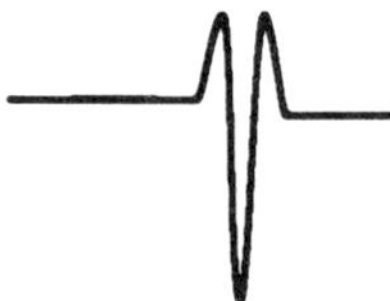

This peak could be easily measured and compared with the second derivative of a suitable captan standard.

Objectives

1. To obtain the first and second derivative of captan using LC with a UV detector.
2. To illustrate how derivative spectrometry can be utilized for quantitation of an analyte separated by LC.

Instrumentation, Solution, and Reagents

Obtain an LC system with a UV detector set at 220 nm and the ability to obtain zeroth, first, and second derivative spectra. Acquire a 25-cm × 4.6-mm I.D. column packed with a cyano-bonded, 5-μm spherical silica (Zorbax CN is available from E.I. DuPont de Nemours, Wilmington, DE) or equivalent.

You will need analytical grade acetonitrile, methylene chloride, petroleum ether, and methanol.

Prepare the mobile phase solvent by adding 450 ml of water to 550 ml of methanol; mix well. Deaerate by bubbling helium gas in the solution just prior to use.

Prepare 20% methylene chloride in petroleum ether by placing 200 ml of methylene chloride into a 1-l volumetric flask and diluting to 1 l with petroleum ether and mixing well.

Acquire 50% methylene chloride + 1.5% acetonitrile in petroleum ether by placing 500 ml of methylene chloride in a 1-l flask, adding 15 ml of acetonitrile, diluting to 1 l with petroleum ether, and mixing well.

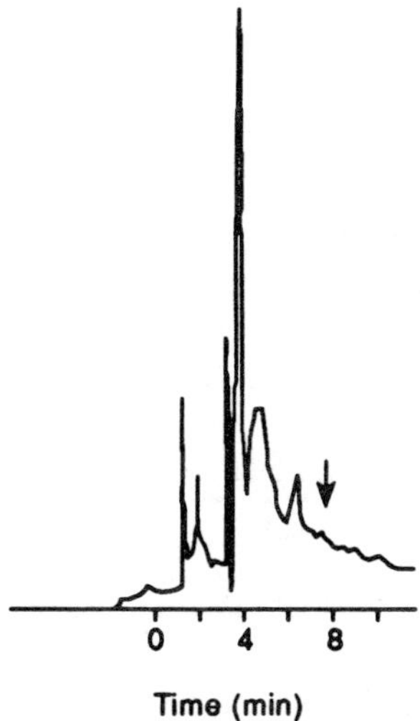

FIGURE 12   UV spectra of captan in strawberry extract.

Procure PR grade Florisil, available from various commercial sources, including Pennsylvania Glass and Sand Corp. (2 Gateway Center, Pittsburgh, PA 15222). Weigh into a flask an amount of Florisil, which when checked with 10 µg of captan, will permit 90 to 110% elution of captan using the 50% methylene chloride + 1.5% acetonitrile in petroleum ether solution. Use a weight of Florisil equal to 85 to 100% of that calculated by Mills' lauric acid number.[115] Heat Florisil portions in full covered flasks at 130°C in an oven overnight to activate. Remove and let cool 1 hour at room temperature and use to pack the chromatographic columns.

Obtain Celite 545, anhydrous granular sodium sulfate, and reagent grade sodium chloride.

Gather the following apparatus: Buchner funnel (12) cm, high-speed blender (Waring® or equivalent), chromatographic columns with Teflon stopcocks and coarse fritted glass plate, and low pressure rotary evaporators.

Obtain captan standard from the EPA or other appropriate commercial outlets.

Prepare captan standard solution at 1 µg/ml in 2-propanol using the following sequence: weigh 1 mg into a 100-ml volumetric flask, dilute to 100 ml, and mix to obtain a 10-µg/ml solution. Take 10 ml of the 10-µg/ml solution into a 100-ml volumetric flask, dilute to volume with 2-propanol, and mix well to give a 1-µg/ml standard solution.

Procedure

Weigh 100 g of strawberries into a blender jar. Fortify by adding 3 µg of captan standard at a 0.03-ppm level. Follow the procedure given in the *Pesticide Analytical Manual,* (Vol. 2, "Captan Method 1," Food and Drug Administration, Washington, D.C., 1975). Evaporate the final solution to dryness in a rotary evaporator under low pressure in a water bath set at 40° to 50°C. Use 2-propanol to wash the solution into a glass tube. Dilute or evaporate to 3 ml to obtain an approximately 3-µg/3 ml or 1-µg/ml solution (assuming 100% recovery).

Prepare a blank solution by taking an unfortified 100-g sample of strawberries through the same procedure.

Comply with the manufacturer's directions and obtain the zeroth, first, and second derivatives of the 1-µg/ml standard solution using a UV detector set at 220 nm. Try injecting 10-µl aliquots and adjust the volume injected and the instrument parameters so as to obtain approximately 50% full-scale deflection.

Obtain the net intensity (the difference between the blank and the pesticide peak) for the standard and the fortified sample. Using all three derivative methods, determine the percentage of recovery using the formula:

$$\% \text{ Recovery of captan} = \frac{[\text{Captan added (ppm)}]}{[\text{Captan found (ppm)}]} \times 100$$

Discussion

Captan exists as a white solid with a melting point of 175°C and has very low solubility in organic solvents. Captan is used for the control of scab, black rot, botrytis, sooty blotch, fly speck, and summer rots on apples; brown rot and leaf spots on stone fruits and almonds; and dead arm, down mildew, and black rot on grapes. It also effectively controls a wide variety of fungus diseases on small fruits, berries, vegetables, and ornamental crops. Captan is used as seed treatment by slurry, dry treatment, and plant box application. The structure of this pesticide is

LC WITH PHOTOCONDUCTIVITY/UV DETECTION USED FOR THE DETERMINATION OF PHENYLUREA PESTICIDES
A REVIEW OF AN ARTICLE BY S.M. WALTERS, B.C. WESTERBY, AND D.M. GILVYDIS[98]

Introduction

High-performance liquid chromatographic (HPLC) separations of 18 phenylurea pesticides were investigated using both reversed- and normal-phase systems. A PCD, which responds selectivity to ionic products formed via postcolumn UV irradiation of photolabile analytes, was connected in tandem with a UV detector permitting serial dual detection of these compounds. The PCD responded selectively to the 13 halogen- and sulfur-containing compounds, whereas the UV detector responses were of the same order of magnitude for all 18 compounds at 250 mn. The HPLC/UV/PCD system was successfully applied to the determination of chloroxuron in strawberries at the official tolerance of 0.5 ppm. The tandem detectors, combined with a choice of columns and chromatographic modes, offers enhanced selectivity for the HPLC analysis of these pesticides as trace contaminants in complex samples.

Analysis by Derivative Spectrometry

The following phenylurea pesticides were separated on a reversed-phase 25-cm × 4.6-mm I.D. Zorbax ODS (6 μm) column at 25°C by a concave gradient from 60 to 75% methanol in water. The numbers on the peaks correspond to the following table:

Peak no.	Pesticide	Peak no.	Pesticide
1	Fenuron	10	Difenoxuron
2	Metoxuron	11	Isoproturon
3	Monuron	12	Diuron
4	Karbitulate	13	Linuron
5	Thiadiazuron	14	Siduron
6	Monolimuron	15	Chlorbromuron
7	Fluometuron	16	Chloroxuron
8	Chlortoluron	17	Diflubenzuron
9	Metobromuron	18	Neburon

Peak pairs 3 to 5, 6 to 7, and 10 to 11 could be better analyzed by obtaining second derivatives. For example, the 3 to 5 pair is illustrative of closely merged peaks. The first and second derivative would resemble:

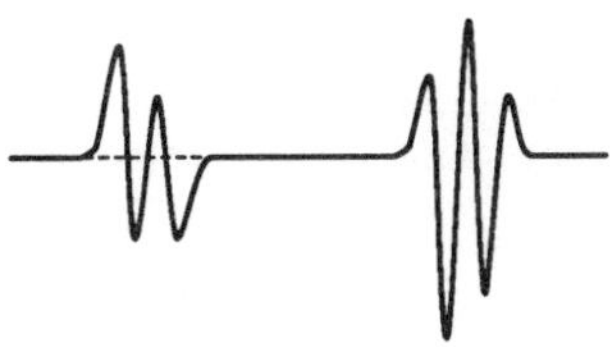

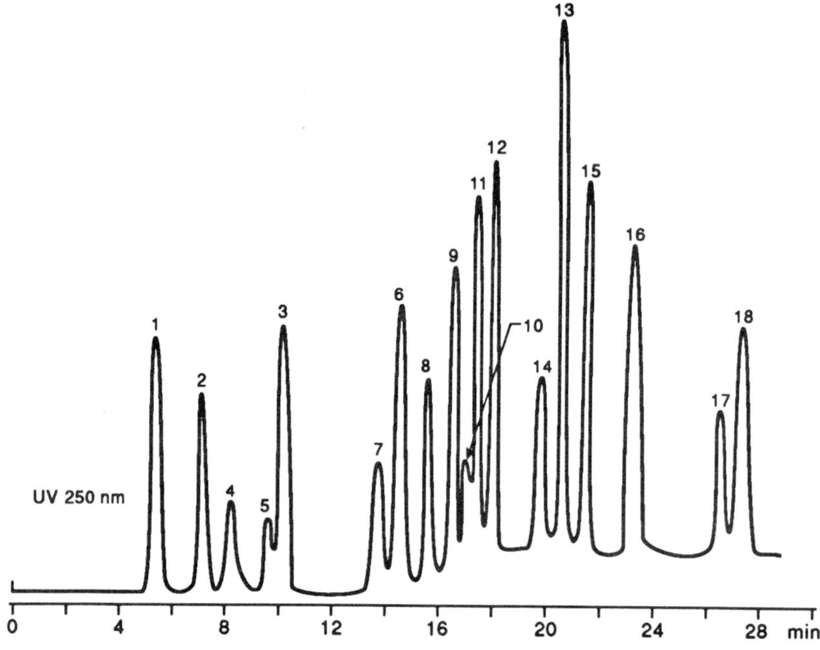

FIGURE 13   Phenylurea pesticides separated by HPLC.

These second derivative peaks could be more easily measured for quantitative analysis than the original UV peaks.

Peak pair 10 to 11 is a similar example of closely merged peaks, and the second derivative spectrum would also resemble the one shown in the previous example.

Peak pair 6 to 7 is an example of slightly merged peaks. The UV peaks could be easily measured for quantitation, but an alternate solution would be to obtain the second derivative spectrum which would probably look like:

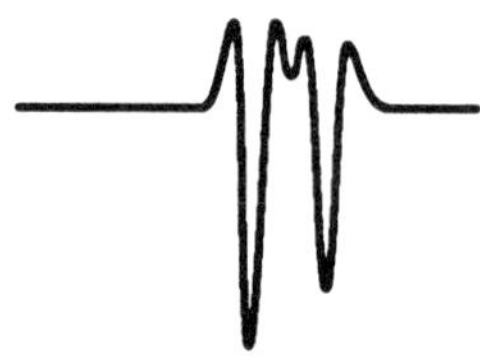

These second derivative peaks could be measured for quantitative analysis very easily and are better separated than the UV peaks themselves.

Objectives

1. To apply derivative techniques to the analysis of certain phenylurea pesticides separated by LC.

Instrumentation, Solutions, and Reagents

Obtain an LC instrument that has a UV detector. The instrument must be capable of obtaining zeroth, first, and second derivative peaks. Acquire a 25-cm × 4.6-mm I.D. Zorbax ODS (6 μm) column and run the LC instrument at 25°C using a concave gradient from 60 to 75% methanol in water.

Prepare a standard solution of monuron, thiadiazuron, monolimuron, fluometuron, difenoxuron, and isoproturon at a concentration of about 10 g/μl by weighing up about 100 mg each of monuron, thiadiazuron, monolimuron, fluometuron, difenoxuron, and isoproturon into a 10-ml volumetric flask. Dilute to 10 ml with 75% methanol in water.

For the sake of illustrating how this method could be used analytically, prepare an artificial sample by making up a second solution containing 9 ng/μl of each of the phenylurea pesticides by weighing up 90 mg each of monuron, thiadiazuron, monolimuron, fluometuron, difenoxuron, and isoproturon into a 10-ml volumetric flask. Dilute to 10 ml using 75% methanol in water. This will be called the "sample".

Procedure

Follow the directions given by the manufacturer for running the LC system and obtain the zeroth, first, and second derivatives of these analytes for the standard, sample, and a blank solution made up of 75% methanol in water.

Measure the net intensity compared to the blank of the zeroth, first, and second derivative peaks of the pesticides in the standard and the sample.

Calculate the percentage of recovery of each of the pesticides in the following way:

$$\text{Pesticide found (ng/}\mu\text{l)} = \frac{\left[\begin{array}{c}\text{Net peak height} \\ \text{of sample}\end{array}\right]}{\left[\begin{array}{c}\text{Net peak height} \\ \text{of standard}\end{array}\right]} \times \left[\begin{array}{c}\text{Conc. of} \\ \text{standard (ng/}\mu\text{l)}\end{array}\right]$$

$$\text{Recovery of pesticide (\%)} = \frac{\left[\begin{array}{c}\text{Pesticide found} \\ \text{in sample (ng/}\mu\text{l)}\end{array}\right]}{\left[\begin{array}{c}\text{Pesticide added} \\ \text{to sample (ng/}\mu\text{l)}\end{array}\right]} \times 100$$

Discussion

The phenylurea pesticides are herbicides that have various functions. Fluometuron is a white crystal with a melting point of 163° to 164°C. It is slightly soluble in water and readily soluble in organic solvents. Fluometuron is effective for preemergence surfaces, preemergence soil incorporated after planting, and postemergence soil weed control of annual grasses and broadleaves. It is also used for weed control in cotton and sugarcane. The structure of fluometuron is

Monolinuron is a solid crystal that is moderately soluble in water. Its melting point is 75° to 78°C. It is used on potatoes, dwarf beans, asparagus, cereals, ornamentals, grapes, berries, ornamental trees, and certain other crops. Its structure is as follows:

Monuron is a white crystal that melts at 174° to 175°C. It has very low solubility in hydrocarbon solvents and water. Monuron is somewhat more soluble in water than is diuron and has a lower coefficient of adsorption. As a soil sterilant, monuron is used on medium to heavy soils. At sterilant dosages it controls a wide range of annual and perennial grasses and broadleaf weeds on noncrop areas. The structure of monuron is given below:

LC WITH PHOTOCONDUCTIVITY/UV DETECTION USED FOR THE DETERMINATION OF OXYFLUORFEN AND OXYFLUORFEN AMINE
A REVIEW OF AN ARTICLE BY MIN ZHOU AND CARL J. MILES[99]

Introduction

Oxyfluorfen and oxyfluorfen amine were determined by LC with PCD/UV. A simple extraction procedure acceptably recovered both analytes from garbanzo beans over a wide range of fortifications (0.05 to 20 ppm; 83 ± 4 for oxyfluorfen; 85 ± 4 for oxyfluorfen amine). Percentage of recovery decreased slightly as the fortification level decreased. Both analytes could be determined simultaneously at a concentration of >0.2 ppm in garbanzo beans. Detection limits were 3 ng for oxyflourfen and 100 ng for oxyfluorfen amine using LC/UV and 12 ng for both oxyfluorfen and oxyfluorfen amine with LC/PCD. Different knitted reaction coils and photoreactors were evaluated. Photoproduct yields and identification were determined by ion chromatography. The LC/PCD method measures oxyfluorfen and oxyfluorfen amine separately and has a shorter analysis time, while the standard method, using GC, measures total residues and is more sensitive.

Analysis by Derivative Spectrometry

Using LC, at the optimum mobile phase of 61% acetonitrile, spectral interferences found in a control garbanzo beans matrix that coelutes with oxyfluorfen and oxyfluorfen amine make quantitative measurements difficult. Figure 14 shows background interference for oxyfluorfen (I) and oxyfluorfen amine (II).

This spectral interference for oxyfluorfen can be overcome, however, using derivative spectroscopy because it allows resolution of two or more spectral bands of close wavelengths and, thus, allows quantitation without separation. For example, if the second derivative was taken of the region of the UV spectra in which the interference and the oxyfluorfen peak occur, the second derivative spectrum might look like the following:

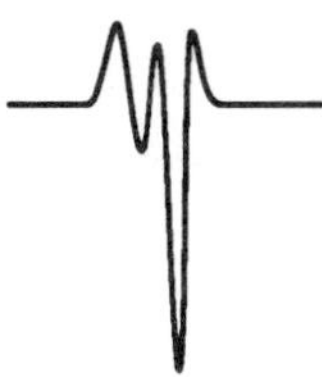

Quantitation of the oxyfluorfen in the sample by comparison of a similarly treated standard would now be very easy.

Because the interference found with oxyfluorfen amine is found underneath the UV peak, derivative spectrometry would not help as only one large second derivative peak would result:

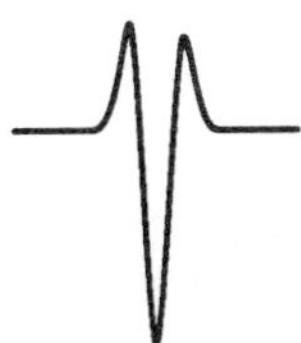

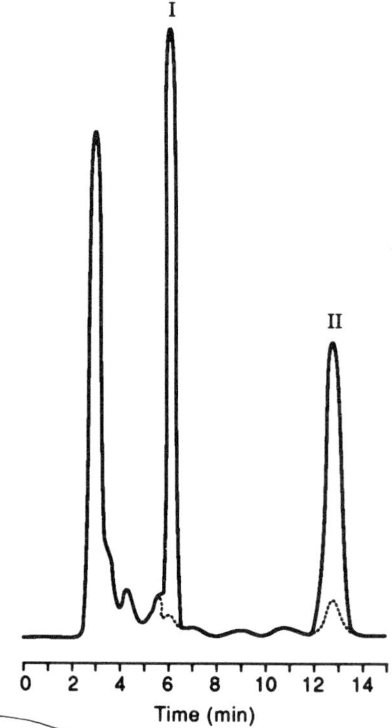

FIGURE 14   LC/UV chromatograms of garbanzo beans containing oxyfluorfen and oxy-fluorfen amine.

Objectives

1. To obtain the zeroth, first, and second derivatives of oxyfluorfen and oxyfluorfen amine using LC.

2. To show how derivative spectrometry using LC can quantitate in the presence of spectral interferences.

Instrumentation, Solutions, and Reagents

Obtain an LC system that is capable of obtaining zeroth, first, and second derivatives. The instrument should be equipped with a 3- $\times$ 30-mm, 3-μm C-18 analytical column with a suitable gradient pump and UV detector. Other reagents include acetonitrile, deionized water, a homogenizer, food chopper, centrifuge, centrifuge bottle, 0.45-μm pore size filter, and a rotary evaporator. These reagents can be obtained from various commercial outlets.

Prepare the mobile phase for LC/UV analysis by preparing 1000 ml of (61 + 39) acetonitrile-deionized water.

Obtain oxyfluorfen and oxyfluorfen amine standard from Rohm & Haas. Prepare a mixed standard solution of oxyfluorfen and oxyfluorfen amine in the following way:

1. Weigh out 1 mg of oxyfluorfen and oxyfluorfen amine into a 100-ml volumetric flask and dilute to 100 ml with (61 + 39) acetonitrile-deionized water to obtain a 10-μg/ml solution.

2. Pipette 10 ml (100 μg) of the mixed standard solution into a 20-ml volumetric flask and dilute to 20 ml with (61 + 39) acetonitrile-deionized water to obtain a 5-μg/ml spiking solution.

Obtain garbanzo beans from a suitable source.

Procedure

Chop garbanzo beans (dried and husked) using dry ice and a Hobart mill. Weigh 10 g chopped sample into Teflon centrifuge tubes. Add 6 ml of 5-μg/ml mixed oxyfluorfen and oxyfluorfen amine standard solution to the 10-g chopped sample to obtain a 3-ppm fortification. Let stand for 1 h at room temperature so that the standard solution can distribute itself. Add 20 ml acetone, homogenize samples for 1 min, centrifuge at high speed for 20 min, and transfer the supernatant to a round-bottomed flask. Evaporate to dryness and reconstitute sample in 2 ml acetonitrile. Add 2 ml of deionized water and filter mixture through a 0.45-μm filter. Inject into the LC system and use the manufacturer's instructions to obtain the zeroth, first, and second derivatives of the analytes found in the standard and sample. The standard used should have a concentration of 5 μg/ml of each of the pesticides. The UV detector should be set at 277 nm. A blank solution prepared by taking 10 g of residue-free garbanzo beans through the same procedure should also be analyzed in the same way and the peak height of the pesticides in the sample and standard should be subtracted from the peak height of the blank.

After examining the zeroth, first, and second derivative spectra, a decision should be made as to which derivative mode would be most suitable for quantitation.

Calculate the percentage of recovery of the oxyfluorfen and oxyfluorfen amine from the garbanzo beans in the following way:

$$\text{Pesticide (ppm)} = \frac{\text{Vol. standard injection}}{\text{Vol. sample injection}} \times \frac{\text{Net peak height of sample}}{\text{Net peak height of standard}} \times \frac{\text{Conc. standard}}{(\mu\text{g/ml})} \times \frac{\text{Final vol. of sample}}{\text{Weight of sample (g)}}$$

$$\text{Recovery of pesticide \%} = \frac{[\text{Pesticide found (ppm)}]}{[\text{Pesticide added (ppm)}]} \times 100$$

Discussion

Oxyfluorfen and oxyfluorfen amine are selective contact herbicides used on a variety of tropical and subtropical crops. Its low mammalian toxicity and its strong sorption to soils and subsequent minimal leaching make it a low environment risk and attractive to users. The structures of oxyfluorfen (I) and oxyfluorfen amine (II) are

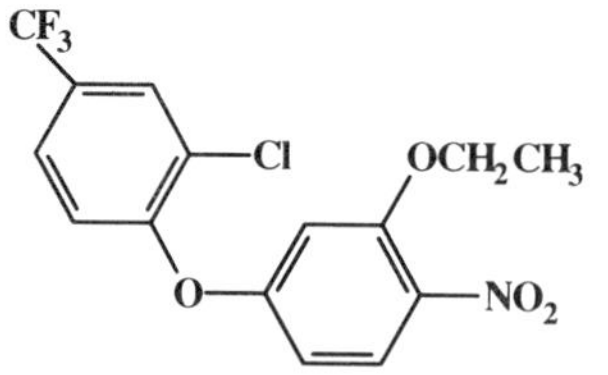

(I 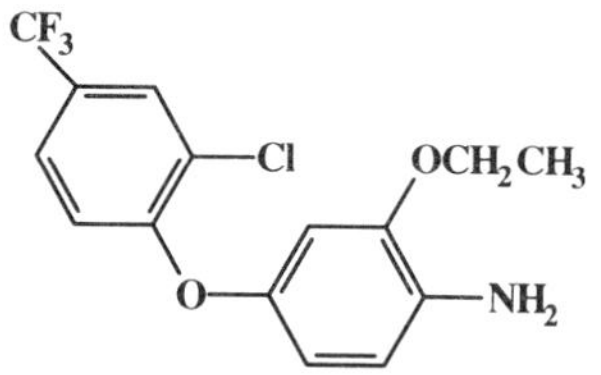)   (II)

LC WITH UV DETECTION USED FOR THE DETERMINATION OF AVERMECTIN Bla AND Blb
A REVIEW OF AN ARTICLE BY JOHAN VUIK[100]

Introduction

A rapid, sensitive, and reliable method is presented for the determination of trace amounts of abamectin in lettuce and cucumber. Abamectin consists of $\geqslant 80\%$ avermectin Bla and $\geqslant 20\%$ avermectin Blb. Vegetables were extracted with ethyl acetate and the extract was purified by solid-phase extraction using Sep-Pak silica cartridges. The purified extracts were analyzed by HPLC with a Zorbax ODS column, 5 μm, and UV detection under isocratic conditions. The method yields recoveries for avermectin Bla and Blb of 76 to 109% in the 0.054 to 0.54 mg/kg range. The limit of detection of the method is 40 μg/kg each of avermectin Bla and Blb in vegetables.

Analysis by Derivative Spectrometry

When cucumber and lettuce extracts are fortified with 540 μg/kg of avermectin Bla and 62.8 μg/kg of avermectin Blb, the resulting chromatograms (Figures 15 and 16) from LC show that avermectin Bla and Blb are very closely merged. The peaks for Blb in particular are very small and would be hard to quantitate.

Second derivative spectrometric measurements have the advantage of intensifying the specificity and precision of the analysis by recording the peaks at selected wavelengths, with the spectral interference eliminated.

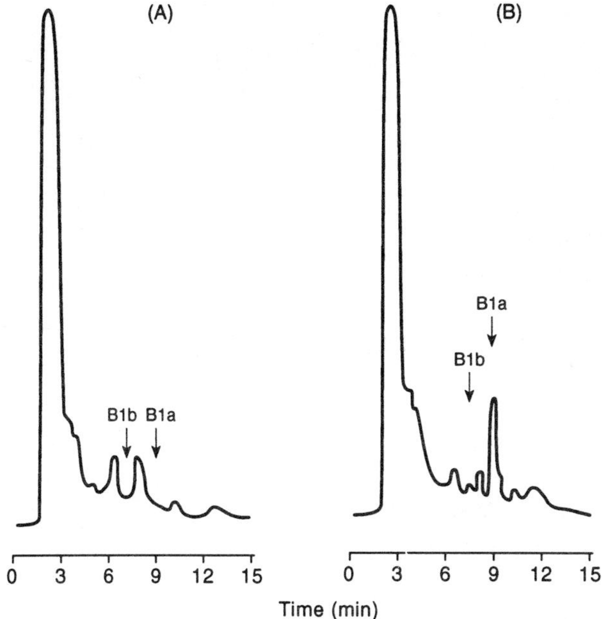

FIGURE 15   LC/UV chromatograms of (A) unfortified cucumber extract and (B) cucumber extract fortified with avermectin B1a and B1b.

The appearance of the second derivative spectra might look like this:

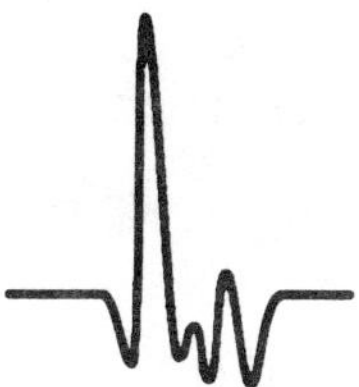

These peaks may be compared to the second derivative of a single standard or a series of standards (if a standard curve was desired), and the quantity of the avermectin B1a and B1b could be easily determined.

Objectives

1. To analyze for abamectin in lettuce using LC and derivative spectrometry.
2. To obtain the zeroth, first, and second derivatives of avermectin B1a and B1b.

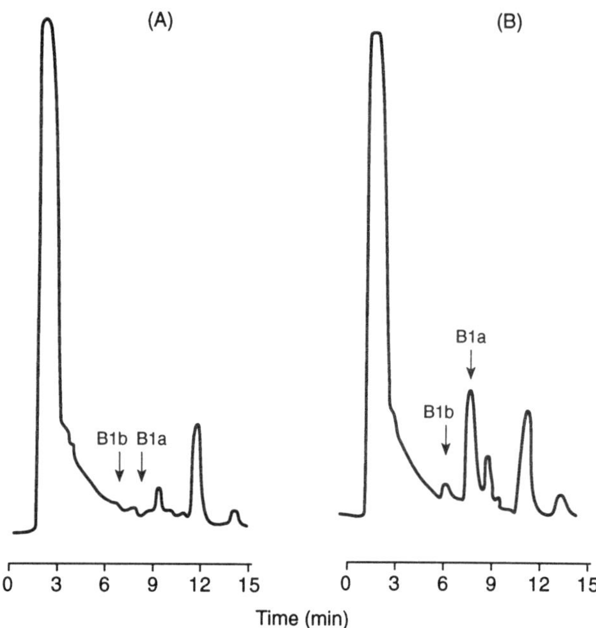

FIGURE 16 LC/UV chromatograms of (A) unfortified lettuce extract and (B) lettuce extract fortified with avermectin B1a and B1b.

Instrumentation, Solutions, and Reagents

Acquire an LC assemblage capable of obtaining zeroth, first, and second derivatives using a UV detector. The column should be an ODS (25 cm × 4.6 mm I.D.) or equivalent, operated at room temperature.

Prepare the mobile solvent by adding 100 ml of water to 900 ml of methanol to obtain a 90:10 methanol-water solution.

Obtain analytical grade ethyl acetate and hexane and several 0.8 g Sep Pak silica cartridges from Waters Associates (34 Maple Street, Milford, MA 01757).

Procure abamectin standards in glycerol containing 1.0074% (w/w) avermectin B1a and 0.125% (w/w) avermectin B1b from Merck, Sharp & Dohme (Haarlem, The Netherlands). Prepare a mixed standard solution containing 54 mg/l avermectin B1a and 6.28 mg/l B1b in methanol by weighing out 5.03 g of the glycerol solution containing 1.074% (w/w) of avermectin B1a and 0.125% (w/w) avermectin B1b. This will contain 54 mg of avermectin B1a and 6.28 mg of avermectin B1b. Transfer to a 1000-ml volumetric flask and dilute to 1000 ml with methanol. Store the stock standard solution in the dark. The standard solution is stable for 4 months.

You will also need a head of lettuce.

Procedure

Chop a suitable sample of lettuce and weigh 10 g into a 250-ml centrifuge tube. Fortify by adding 500 µl of abamectin standard (54.0 mg/l avermectin Bla and 6.28 mg/l avermectin Blb). Add 100 ml of ethyl acetate and shake vigorously for about 5 min. Centrifuge at high speed for 1 min. Transfer 50 ml of the supernatant to a round-bottom flask and evaporate to dryness on a rotary evaporator using a water bath set at about 40°C under reduced pressure. Dissolve the residue in 2 ml of ethyl acetate and 3 ml of hexane. Mix well. Prewash a 0.8-g Sep Pak silica cartridge with 8 ml of 40:60 ethyl acetate-hexane and discard the eluate. Add the contents of the flask to the silica solid phase extraction (SPE) cartridge. Rinse the flask twice with 1 ml of 40% ethyl acetate in hexane. Elute the pesticides from the silica SPE with 5 ml of ethyl acetate-methanol 50:50. Evaporate the eluate to dryness on a rotary evaporator in a water bath at 40°C under reduced pressure. Dissolve the residue in 1 ml of this solution.

Prepare a blank solution by taking 50 g of unfortified lettuce through the same procedure.

Optimize the LC conditions and obtain the zeroth, first, and second derivatives of the pesticides in the standard and the sample by following the manufacturer's directions. A blank solution should also be analyzed. About 25 µl should be injected into the LC/UV detector.

Obtain the net intensity (net difference between the blank and the pesticide peaks) for the standard and the fortified sample and calculate the percentage of recovery in the following way:

$$\text{Avermectin Bla or Blb (mg)} = \left[\frac{\text{Vol. standard injected}}{\text{Vol. sample injected}}\right]\left[\frac{\text{Net peak height of sample}}{\text{Net peak height of standard}}\right] \times \left[\text{Std. conc. (mg/ml)}\right] \times \left[\text{Final vol. sample (ml)}\right]$$

$$\text{Recovery of avermectin Bla or Blb (\%)} = \frac{\left[\text{Pesticide found (mg)}\right]}{\left[\text{Pesticide added (mg)}\right]} \times 100$$

Discussion

The avermectins are a family of pesticide agents that are produced by the actinomycete *Streptomyces avermitilis*. Abamectin is the commercial product that was developed as an insecticide. It has been used against fire ants.

Abamectin consists of >80% avermectin B1a and <20% avermectin B1b. These compounds have the following structures:

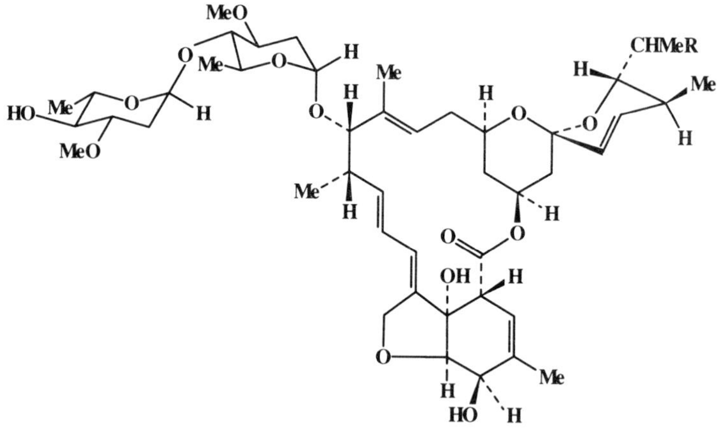

Structures of the two major components of abamectin are B1a (R = C_2H_5) ≥80%, and B1b (R = CH_3) ≤20%. Me = CH_3.

LC WITH AMPEROMETRIC DETECTION USED FOR THE DETERMINATION OF 3-PHENYL-4-HYDROXY-6-CHLORPYRADAZINE
A REVIEW OF AN ARTICLE BY ANTON PACHINGER, ELISABETH EISNER, HELMUT BEGUTTER, AND HUBERT KLUS[101]

Introduction

The main metabolite of the herbicide pyridate is 3-phenyl-4-hydroxy-6-chlorpyridazine (CL9673). An HPLC method with amperometric detection is described for determining CL9673 at residue levels in water samples. Sample preconcentration is performed by passage through a C-18 extraction cartridge. A recovery study using tap water samples spiked with CL9673 at a concentration of 0.1 µg/l showed a recovery of 84.8% (C.V. 6.2%). The method is suitable for the determination of CL9673 in drinking water and groundwater.

Analysis by Derivative Spectrometry

A chromatogram of Vienna tap water fortified with 0.1 µg/l CL9673 is shown in Figure 17. The CL9673 is eluted from the column with a retention time of 8.05 min. Although there is sufficient separation of the

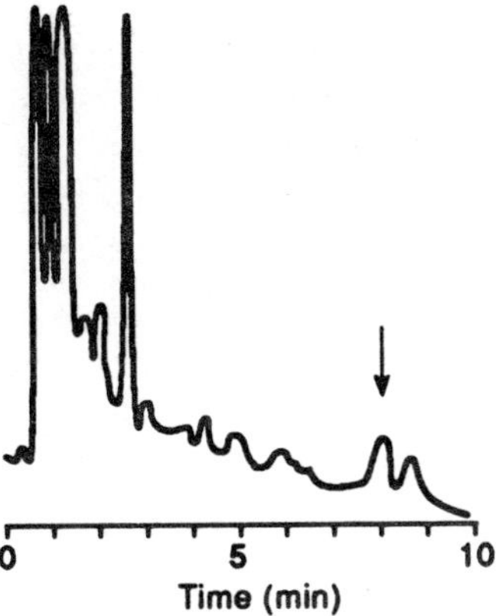

Time (min)

FIGURE 17 HPLC chromatogram of Vienna tap water fortified with 0.1 μg/l CL9673.

CL9673 peak, the CL9673 is slightly merged with an interference peak. The analysis of some substances with overlapping or near-overlapping spectral bands presents some hardships, but via second derivative spectroscopy, these slightly merged absorption bands can be separated. For example, the second derivative of the slightly merged absorption peaks might appear as:

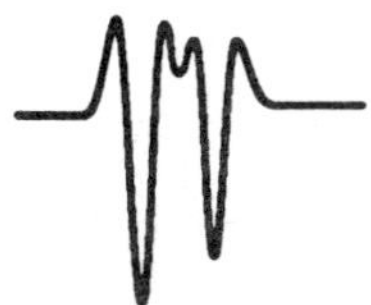

This second derivative spectra presents an alternative analysis for CL9673.

Objectives

1. To obtain the zeroth, first, and second derivative spectra of CL9673.
2. To analyze tap water fortified with CL9673 by separating the analyte using LC and obtaining the zeroth, first, and second derivative spectra.

Instrumentation

Procure an LC system with a programmable electrochemical detector in the emperometric mode and equipped to obtain zeroth, first, and second derivative spectra. The column selected for use must be a reversed-phase 5-μm ODS-Hypersil column (100 × 2 mm I.D.) or equivalent.

Obtain analytical grade acetic acid, acetone, methanol, and sodium chloride.

Prepare the mobile solvent in the following way.

- Add 150 ml methanol to 850 ml water in a graduate and mix well.
    1. Add sodium chloride so as to make a 3.5-mmol sodium chloride solution by weighing out 0.205 g sodium chloride and placing the reagent in a 100-ml volumetric flask.
    2. Add 3 ml acetic acid, 150 ml methanol, and dilute to 1000 ml with distilled water.

Thus, a mobile solvent containing methanol-water (15:85) which is 0.0035 M in sodium chloride and 0.3% in acetic acid is prepared.

Obtain CL9673 standard from Chemie Linz (Linz, Austria) or any other suitable commercial source. Prepare a standard stock solution of 1 mg/ml by weighing 100 mg into a 100-ml volumetric flask. Dilute to 100 ml with methanol to obtain a 1-mg/ml standard solution.

Prepare diluted standard solutions by placing 1 ml (1 mg) of the stock solution into a 500-ml flask and dilute to volume with water and mix well. This gives a 2-μg/ml solution.

Prepare solutions for a standard curve by proceeding through the following steps:

1. Add 2 and 1 ml of the 2-μg/ml standard solution into 10-ml volumetric flasks. Dilute to 10 ml with distilled water and mix well. Thus, 0.4- and 0.2-μg/ml standard solutions are prepared.
2. Add 2.0 and 1.0 ml of the 2-μg/ml solution into 100-ml volumetric flasks. Dilute to 100 ml with distilled water to obtain 0.04- and 0.02-μg/ml standard solutions.
3. Add 2 ml of the 2-μg/ml standard solution to a 1000-ml volumetric flask. Dilute to 1000 ml with distilled water and mix well to obtain a 0.004-μg/ml solution.

Obtain tap water.

Procedure

Take 300 ml of tap water and fortify by adding 5 ml of the 0.02-μg/ml standard solution (0.1 μg) to the tap water. Adjust the water to pH 2 with 6 M HCl using a pH meter. Add 5 ml of methanol. Prewash a 1-g C-18 SPE cartridge with 5 ml methanol followed by 5 ml of distilled water. Discard the prewash. Pass the water through the C-18 using a vacuum to obtain a 10-ml/min rate and discard the eluate. Let the SPE air dry by pulling air through the cartridge for about 30 min. After this, elute the pesticide with 6 ml of acetone. Concentrate the eluate to about 0.2 ml under nitrogen. Dilute to 0.5 ml using a microliter syringe. Filter through a 0.5-μm membrane filter.

Prepare a blank solution by taking unfortified tap water through the same procedure.

Follow the manufacturer's directions and obtain the zeroth, first, and second derivative of each of the standard solutions used to make the standard curve (0.4, 0.2, 0.04, 0.02, and 0.004 μg/ml). Also, obtain the zeroth, first, and second derivatives of the blank and the fortified sample. Aliquots of 25 μl should be injected at the start and the response used as a guide to the desired injection volume.

Obtain the net intensity (the difference between the blank and the pesticide peak) for the standards and the fortified sample. Plot a standard curve and determine the micrograms per milliliters of the pesticide in the fortified sample. Determine the percentage of recovery in the following way:

$$\text{Recovery of CL9673 (\%)} = \frac{\left[\begin{array}{c} \text{CL9673} \\ \text{found } (\mu\text{g/ml}) \end{array} \right]}{\left[\begin{array}{c} \text{CL9673} \\ \text{added } (\mu\text{g/ml}) \end{array} \right]} \times 100$$

Discussion

The herbicide pyridate {*0*-[3-phenyl-6-chlorpyridazinyl-(4)]-*S*(*n*-octyl)-thiocarbonate} is used in the agricultural management of cereals, maize, and rape. It is also used in the postemergence of weeds such as Florida beggarweed, pigweed, lambsquarters, kochia, cocklebur, sicklepod, velvetleaf, morning glories, and triazine-resistant weeds. This herbicide is a brown, oily, nonvolatile liquid, with a melting point of 20° to 25°C. The metabolite of this compound is 3-phenyl-4-hydroxy-6-chlorpyridazine (CL9673). The structure of CL9673 is given below:

CONCLUSION

In this section, I have attempted to reveal an alternative method to analyze for pesticides in cases in which the analyte being quantitized has absorbance or fluorescence peaks that either overlap or nearly overlap interfering peaks when run by LC. Thus, derivative spectrometry could be a supplement or an alternative to LC in which UV, visible, or fluorescence spectrometry is used. Derivative spectroscopy is not applicable in all cases, but in many it could be useful. At the present time, to the best of the author's knowledge, derivative spectroscopy has not been applied to pesticides. Future researchers should look into this technique.

Section III

History of Derivative Spectroscopy and Fluorometry

FOREWORD

In most books of this type, little mention is made of the history of a particular technique. In most cases, a brief summary of the history is given along with suitable references. In some cases, an author will claim that the theory behind the technique is beyond the scope of the book. I think this is a disservice to the reader who may or may not be able to understand the theoretical background. For this reason, I am recounting the history of derivative spectrometry and derivative luminescence using the theoretical equations associated with a particular author. Each author is identified with: (1) the type of instrument used; (2) the theory of the instrument, including equations; and (3) what, if anything, was analyzed. Conclusions are drawn at the end of each discussion.

Because so many contributed to the development of this technique, only a limited number of authors' works are reviewed in order to obtain a broad overview.

ARTHUR T. GIESE AND C. STACEY FRENCH[102]
DERIVATIVE SPECTROMETRY

Derivative spectrometry was introduced in 1952 by French and elaborated on by Giese. The authors state:

> In spectroscopic analysis, particularly of biological and biochemical materials, it is often important to detect and to determine quantitatively the position of a weak absorption band overlaid or obscured by another absorption band. The difficulties of measuring such minor bands are the same whether the small band and the one that obscures it belong

to different substances or whether they are component bands of a complex absorption spectrum of a single substance. To supplement present methods of absorption spectrophotometry, a new type of spectroscopic measurement has been proposed which may be extremely useful in certain cases. This technique, designated derivative spectrometry, is simply to record the slope $dT/d\lambda$ of the percent transmission curve as well as the extinction or the transmission curve itself. The small details of complex band shape upon which wavelength position of maxima and qualitative identification of mixed substance depend are shown much more clearly in derivative measurements. The actual height of the integral curve, as used for concentration determination, is, however, better measured in the usual way than by integration of the differential curve.

Derivative spectrophotometry yields a plot of the slope of the transmission curve of one sample. This paper is an exploration of the behavior of the derivatives of hypothetical overlapping absorption bands. It shows the inherent advantages in accuracy that can be gained by using derivative measurements in resolving overlapping bands.

Type of Instrument Used

The derivative is produced in a form such that all of the quantities can be set up as curves or voltage outputs on the graphical computer, or can be arranged as columns of numbers to be dealt with on a desk calculator. For the curve most frequently used below having a transmission minimum of 0.40, the values of the constants are $c = 1/\sqrt{2\pi}$, $b = 1/2$.

Theory Applied to the Instrument Used

The basic band shape decided upon was probability curves, which can be said to correspond accurately to measured absorption curves of various substances. In this paper, the x axis was called *wavelength,* although it could be equally well called *frequency,* and frequency scales usually do give better fits of probability curves to spectral absorption than do wavelength scales.

Probability curves of various heights, widths, and separations were added to form hypothetical absorption curves. These curves were then converted into terms of transmission; its first derivative was the form in which the records were expressed. Using the mathematics of this conversion, it was found that the equation of transmission curve formed in this way can be differentiated. The derivative curves must be synthesized according to mathematical equations because the mechanical differentiation of the curves was neither easy nor precise.

Derivation of Equations Used to Plot the Derivative Curves

1. Transmission, $T =$ Intensity of transmitted light/Intensity of incident light
Percentage of transmission $= 100\,T$

2. Extinction or optical density, $E = \log_{10} 1/T$
3. $T = 10^{-E} = e^{-2.303E}$

The individual hypothetical absorption bands were constructed from the equation:

4. $E = ce^{-b(x-\bar{x})^2}$

Where $\bar{x}$ is the wavelength of the band center, x represents the wavelength and b and c are constants controlling respectively, the width and height of the curve. When $c = 1/\sqrt{2\pi}$ and $b = \frac{1}{2}s^2$ (s being the standard derivation), this is the equation for the normal probability curve. When two of these hypothetical absorption curves overlap, the following equations apply:

5. $E = E_1 + E_2$
6. $E = c_1 e^{-b(x-\bar{x})^2} + c e^{-b_2(x-\bar{x}_2)^2}$
7. $dE = c_1[-2b_1(x - \bar{x}_1)]e^{-b_1(x-\bar{x}_1)^2} + c_2[-2b_2(x - \bar{x}_2)]e^{-b_2(x-\bar{x}_2)^2}dx$
8. $dE = -2[b_1E_1(x - \bar{x}) + b_2E_2(x - \bar{x}_2)]dx$ (4) 1n (8)
9. $dT = -2.3e^{-2.3E} dE$ *from (3)*
10. $dT = -2.3TdE$ (3) 1n (9)
11. $dT/dx = 2 \times 2.3T[b_1E_1(x - \bar{x}_1) + b_2E_2(x - \bar{x}_2)]$ (8) 1n (10)

What Was Analyzed

Nothing was analyzed. A graphic study was made of the behavior of the derivative curves for the overlapping absorption bands of differing heights, widths, and separation intervals, and the derivatives were compared with the corresponding extinction and percentage of transmission curves. The curves presented were intended as illustrations of the possibilities inherent in the derivative spectrophotometry and for the identification of the extinction or transmission curve types from measured derivatives. A machine was under development in that laboratory to plot the derivative.

Conclusion

In the measurement of absorption spectra, it was difficult to detect low-intensity bands which were overlapped by bands of higher intensity. Their detection was greatly facilitated by measuring the first derivative of the transmission curve with respect to wavelength.

G. L. COLLIER AND F. SINGLETON[103]
INFRARED ABSORPTION DERIVATIVE SPECTROMETRY

Collier and Singleton (1956) applied the technique to infrared (IR) absorption by taking the second derivative of the spectrum. Their instrument utilized an analog computer in order to obtain derivative spectra.

Type of Instrument Used

The spectrometer used as a basic unit in this instrument was the Grubb Parsons S3a, modified to meet the new requirements of derivative spectroscopy. The rotating vane light interrupter was found to be responsible for the high noise level in the derivative due to short-term instability of the flicker frequency. It was consequently replaced by a Goodman's vibrator, fitted with a small blade to interrupt the light beam at a primary focus, and controlled by an oscillator tuned to the amplifier frequency. New power supplies were provided for the amplifier, series valve stabilizers replaced the neon stabilizers, and the valve heaters were connected to a stabilized supply. An amplifier frequency change to 19 c/s was carried out to raise the frequency of any beats between the mains and the signal frequency.

The synchronous motor driving the scanning system was replaced by a direct current (DC) motor generator operating as a velodyne. This ensured a constant and reproducible rate of scan using the repetitive scan facility. Different speeds were obtained by switching the bias conditions of the controller.

A further addition was made to the spectrometer in a high-temperature vapor cell, which was evacuated before adding the sample. Thus, any material that may be volatized at low pressures and temperatures below 250°C may be introduced as a known weight directly into the cell in 2 or 3 min. Moreover, the preparation and handling of solutions was avoided and there was no solvent background absorption which may necessitate a measurement of $(d^2\log I_o)/(dt^2)$.

The required function of the intensity was derived from the spectrometer output by means of an analog computer.

The alternating current (AC) signal from the spectrometer amplifier is rectified and smoothed by homodyne and integrator circuits and a voltage proportional to $\log I$ is then obtained by making use of the characteristics of the metal rectifiers at low current levels. An alternative linear path was provided by switching a resistance into circuit in place of the rectifier. A DC amplifier incorporating a DC negative feedback loop to stabilize the gain passed the signal onto the first differentiator and a cathode follower transferred the first derivative signal to the second differentiator.

These two differentiators had a performance equivalent to that of a conventional CR differentiator with a feedback amplifier of gain 15,000.

Because the signal-to-noise ratio was improved by the use of high scanning speeds, all quantitative work was carried out using the highest permissible rate. Under these circumstances, a reading-out system was necessary to extract the information required from the rapidly changing output voltage. The reading-out system comprised four units: the reading circuit, selector, averaging system, and the bridge unit. Selection of the peak whose maximum amplitude was to be measured, or the wavelength at which the output voltage was required, was done by means of the selector unit. The reading circuit charges a condenser to this voltage while the averaging system collected six successive measurements and averaged them to eliminate slight variations due to noise. The bridge unit made the final answer available on a long scale meter.

The system as a whole was controlled by the scanning mechanism of the spectrometer to which was fitted the potentiometer that provided a time base for the cathode ray tube. During the scanning period, a raising voltage obtained from this potentiometer triggered voltage discriminating circuits in the selector and operated a relay that switched a condenser in and out of the reading circuit. The exact moment during the scan in which this switching took place was determined by the position of the selector controls and the correct dial readings were found by a high-speed recording of the output; a marker pulse was fed to a second channel on the recorder to indicate the interval over which the condenser was the charging circuit. If a peak amplitude was to be measured, the marker was set to straddle the peak required. If a measurement at a given wavelength was to be made, the marker was set to end at that wavelength.

Qualitative records of derivative spectra could be obtained using a recording ammeter provided that the scanning speed was low enough for the pen to respond. Spectra so produced were readily reproducible under identical conditions, but allowances had to be made on the wavelength scale for the delay caused by filter circuits. Some distortion was likely to occur when the signal dropped rapidly toward zero because the movement of the pen depended entirely on the restoring force. High-speed recorders produced a good record of the form of the spectra and were used with scanning speeds, but both types of recorders failed to give a deflection that was accurately proportional to the amplitude of the signal. If quantitative spectra of a group of substances were required, a position plot of peaks and crossovers was made from a high-speed recorder trace, correcting for the delay that was determined from the position of one or more prominent peaks. The peak heights were then measured under standard conditions using the reading system. The satellites were dealt with by using the fixed wavelength setting which measured voltages negative to the zero level. A final plot was then made showing position and relative amplitudes of all the features of the derivative spectrum.

Theory Applied to the Instrument Used (Including Equations)

A single-beam IR spectrometer produced an output voltage that represented the intensity of light falling on the detector and was usually provided with a means of scanning the spectrum so that a record of intensity against wavelength was obtained. If, when the spectrometer was scanning, the DC output voltage was differentiated with respect to time, the derivative of the light intensity with respect to time was recorded. By employing a constant wavelength scan, the derivative with respect to time was proportional to the derivative with respect to wavelength.

While the voltage from the spectrometer amplifier was always positive, the first derivative was positive or negative depending on whether the intensity was decreasing or increasing. In the wavelength interval occupied by an absorption band, the first derivative of the band has two peaks, one corresponding to the maximum rate of fall of intensity and the other to the maximum rate of rise of intensity. Thus, two peaks were obtained in the derivative record in the space of a single absorption band and were, therefore, narrower and sharper than the original absorption band as it was traced by the spectrometer. Because the first derivative was zero at the maxima and minima of the absorption band, it was desirable to use even derivatives that would possess peaks that corresponded to absorption bands. The second derivative spectrum had three peaks per absorption band; one derivative peak corresponded closely in position to the absorption maxima and the other two were satellites of opposite signs on either side of it. The fourth derivative had five peaks per one absorption band, four of which were satellites, two positive and two negative. The second derivative was found to be the most useful because sharp peaks were obtained without an excessive amount of satellite peaks.

As any reversal of the slope of the intensity-wavelength curve resulted in a change in the sign of the first derivative, two absorption bands, which almost completely overlapped but just resolved, gave rise to a first differential that had two positive and two negative peaks and a second derivation that showed the two bands separated by a satellite. An inflection in the intensity wavelength curve gave rise to a minimum in the first derivative and a distinct peak in the second derivative corresponding in position to the inflection. A peak in the second derivative appeared even when the inflection was only a single change in slope in the intensity curve and so small a change from a smooth curve as to produce only an inflection in the first derivative.

The optical density of a material is expressed as $\Delta = clk$, where c is the concentration, l the path length, and k the appropriate extinction coefficient. Differentiating twice with respect to λ gives $d^2\Delta/d\lambda^2 = cl d^2k/d\lambda^2$, again showing proportionality to concentration. Also, according to Beer's law, the total optical density Δ of a mixture of components 1, 2, 3, etc. is equal to $\Delta_1 + \Delta_2 + \Delta_3 +$, etc., which are the separate optical densities. Similarly, $d^2\Delta/d\lambda^2 = d^2\Delta/d\lambda^2 + d^2\Delta/d\lambda + d^2\Delta/d\lambda^2$, etc. The two quantities $d^2k/d\lambda^2$ and $d^2\Delta/d\lambda^2$ were described as sharpness coefficient and sharpness, respectively (similar to the extinction coefficient and the extinction). The sharpness with respect to time was denoted by $\psi(t)$.

$$\Delta = \log I_0/I = \log I_0 - \log I$$

and

$$\psi(t) = d^2\Delta/dt^2 = \frac{d^2\log I_0}{dt^2} - \frac{d^2\log I}{dt^2}$$

Since the background level I changed very slowly as the spectrum was scanned, the rate of change of the slope of I was quite small compared to the rate of change of slope intensity I in the region of the absorption band. Thus, the output $d^2\log I/dt^2$ of a single beam derivative spectrometer was taken to be equal to $d^2\Delta/dt^2$ and no measurement of I_0 was generally required.

What Was Analyzed

Analysis was made on *o*-, *m*-, and *p*-cresols in a mixture in which all three were present. Phenol was also analyzed in a mixture of all three cresols.

Conclusion

Derivative IR spectra were obtained by means of an analog computer and utilized in qualitative and quantitative analyses. The second derivative with respect to time of the light intensity was used to detect and fix the position of minor changes of absorption, while the second derivative of the logarithm of the light intensity approximated the second derivative of optical density and is a quantitative function of the absorbing material. The value of the derivative function was found to be zero or very small over large intervals of the wavelength range, and analysis for a particular ingredient in complex mixtures could very often be made by a single measurement of the magnitude of the quantitative derivative function at an appropriate wavelength without interference from other components.

G. BONFIGLIOLI AND P. BROVETTO[104]
SELF-MODULATING DERIVATIVE SPECTROMETRY

Bonfiglioli and Brovetto constructed a self-modulating derivative optical spectrometer that employed a vibrating mirror to modulate the image of the spectrum. They showed that by modulating the spectrum spatially and at the appropriate frequency, the derivative of the IR spectrum was obtained. The signal-to-noise ratio of the derivative signal increased slightly as compared to the modulated signal.

Type of Instrument Used

The spectroscopic technique rested upon the use of a simple self-modulating derivative spectrometer that was built easily.

Theory Applied to the Instrument Used

Beginning with the case of absorption spectroscopy, the authors considered an optical monochromator and assumed that a vibrating mirror put before the exit slit modulated the wavelength of light. This wavelength-modulated light crossed a specimen having a wavelength-dependent absorption. The authors assumed that they were making use of an idealized source of light with constant spectral density, a light detector with constant spectral sensitivity, and a monochromator with constant dispersion. Under such conditions it was clear that the electric signal obtained from the light detector contained a modulation whose depth was proportional to the derivative of the specimen absorption vs. the wavelength. Also, phase rectification of the signal gave rise to a spectral record with many favorable features.

What Was Analyzed

Nothing was analyzed.

Conclusion

This paper presented the theoretical principles underlying a new technique of optical spectroscopy, characterized by the fact that the record of a spectrum, differentiated and efficiently purified of noise, was obtained in a direct and simple way.

I. ARAMU AND A. RUCCI[105]
SELF-MODULATING DERIVATIVE DENSITOMETER

Aramu and Rucci also developed theoretical intensity expressions for derivative spectrometers.

Type of Instrument Used

The authors devised and built a self-modulating derivative densitometer and describe the features. A rough prototype was assembled. The authors took an enlarged image ($5\times$) of a spectrum, which was recorded on a photographic plate, and projected it on the plane of a micrometric slit after reflection on a mirror surface. This mirror was brought into vibration by a bifilar galvonometer which was powered by an oscillator. This causes the spectrum image to oscillate transversely to the slit edges.

A low-noise phototube was placed behind the slit and sent its electrical signal to the phase rectifier and, after amplification, to a recorder. An electric device provided an automatic scanning of several spectra recorded on the same plate. A 1000-Hz vibration frequency was used.

Theory Applied to the Instrument Used (Including Equations)

The light that falls on the phototube was given by the following equation:

$$\phi = h \int_{x_1}^{x_2} \varphi r(x) dx \tag{1}$$

where h was the height of the slit, x the abscissa across the spectrum (normal to its edges, the coordinates of which were x_1, x_2), and φ was the light flux that fell on the photographic plate, whose transmission factor was r. When the mirror vibrated, the r function became $r(x + k\sin\omega t)$. In this way, Equation 1 yielded

$$\phi = h \int_{x_1}^{x_2} \varphi r(x + K\sin\omega t) dx \tag{2}$$

where ω is the angular frequency at which the mirror vibrated. Then, the electrical signal of the phototube was

$$i = \sigma h \int_{x_1}^{x_2} \varphi r(x + K\sin\omega t) dx \tag{3}$$

where σ was a proportionality constant depending on the phototube sensitivity. When the slit width $x_2 - x_1$ was sufficiently small this yielded $i = \sigma h(x_2 - x_1)\varphi r(x_0 + K\sin\omega t)$. This is the abscissa at the center of the slit.

What Was Analyzed

No analysis was made.

Conclusion

The main features of a self-modulating derivative densitometer were reported. By comparing both derivative and nonderivative analysis of a spectrum conventionally recorded on a spectrographic plate, one could see that its detectability was increased by a factor much greater than one. These experimental results were consistent with the gain to be expected by means of optical derivative spectroscopy.

G. BONFIGLIOLI, P. BROVETTO, G. BUSCA, S. LEVIALDI, G. PALMIERE, AND E. WANKE[106]
SELF-MODULATING DERIVATIVE OPTICAL SPECTROMETER

Bonfiglioli et al. completely described a self-modulating derivative optical spectrometer (SMODOS).

Type of Instrument

This paper described a new kind of absorption spectrometer. The instrument enabled the authors to record the derivative of the optical density of an absorbing specimen through a self-modulation operation. The SMODOS that was described was a first-edition prototype and basically consisted of a monochromator that operated in the visible range, modified by the interposition of a vibrating mirror in the light path in such a way that it emitted a wavelength-modulated beam.

An old Hilger spectroscope equipped with a Pellin-Broca constant deviation glass prism was converted to monochromator operation. On the optical path, after the objective lens and about 18 cm before the exit slit, a plane mirror (25 × 45 mm), vibrating around an axis parallel to the prism edge, was interposed.

The electromechanical driving system was derived from a special loudspeaker that was capable of following without distortion frequencies much higher than those employed (200 to 1500 c/s). This setup originated the wavelength modulation indicated in Equation 1. The light coming from the exit slit fell on a beam splitter which split the beam in two; one part went directly to the monitoring phototube, while the other went to another similar phototube after passing through the specimen.

The phototubes were two RCA 931-A photomultipliers that were accurately matched. The beam splitter was achromatic and consisted of a total reflection prism intercepting a fraction of the incoming beam. A micrometric screw permitted regulation of the relative intensity of the two beams so obtained. In the optical path, just after the specimen holder, through a T junction, a sliding eyepiece enabled the authors to check the correct focus and alignment of the entire setup. Suitable shutters were placed before each phototube, both for protection and calibration purposes. The wavelength scanning drum was motor driven and able to operate at the most suitable speed for recording the particular spectrum under study. The resolving power of this instrument was good enough to separate the sodium doublet; however, because of the poor luminosity (numerical aperture 0.1), the authors usually worked under worse resolution conditions, although they were able to separate the mercury doublet (5769 to 5790 Å).

Theory Applied to the Instrument Used (Including Equations)

In the equations to follow, f was the function that determined the correspondence between the wavelength and the abscissa x across the exit plane. The modulation law was written by Taylor's expansion as

$$\lambda(t,\delta = f(x + k\sin\omega t) = f(x) + (df/dx)k\sin\omega t + (d^2f/dx^2)(k^2/s)\sin\omega t + \ldots$$

where the derivatives were taken at $t = 0$, ω stands for the angular frequency (1 kilocycle per second) at which the mirror vibrates, and k was a constant depending upon the vibration amplitude.

Calling x_0 the abscissa of the middle point of the slit, the wavelength λ of the outgoing beam, for the small values of k_1, was given by the equation:

$$\lambda(t,x_0) = f(x_0 + k\sin\omega t) = (\lambda_0 + df/fx)k\sin\omega t$$

The symbol λ_0 indicated the quantity $f(x_0)$.

The authors then considered the output current of a phototube illuminated by a fraction of a monochromator light obtained through a neutral beam splitter and filtered through an absorbing specimen. They saw that this current was actually amplitude modulated at a rather small depth. Moreover, from a second phototube directly illuminated by the second fraction of the monochromator beam, they obtained a monitoring signal. This was mixed with the specimen signal in an electronic circuit that gave an output proportional to the logarithm of the ratio of the two photomultiplier currents. For the sake of brevity, they referred to this circuit as LORAC (an acronym for logarithmic ratio circuit). The LORAC output, after rectification and purification from the noise found in a lock-on amplifier tuned at the angular frequency ω, enabled them to record a quantity $S(\lambda_0)$ proportional to the derivative of the specimen optical density $\delta(\lambda)^*$ vs. the wavelength. This was expressed as:

$$s(\lambda_0) = k(df/dx_0)(d\delta/d\lambda)\lambda_0$$

In this way the SMODOS operated similarly to an electron spin resonance spectrometer, making it possible to transpose to the optical case the well-known good performance of the latter.

What was Analyzed

Derivative spectra of solutions of $(P + Nd)(NO_3)$ in $Cu(NO_3)_2$ solution were obtained.

Conclusion

The authors demonstrated that SMODOS behavior closely followed the predictions of its theory in the sense that all the relevant parameters actually could be controlled to a good degree of accuracy. Moreover, the practical utility of derivative spectroscopy could be considered proven.

A. PERREGAUX AND G. ASCARELLI[107]
WAVELENGTH MODULATION OF OPTICAL SPECTRA

Type of Instrument Used

An oscillator that could be phase locked, e.g., the Hewlett Packard 3300A or 200 CD oscillators, produced a sinusoidal signal at frequency f, which was amplified by a high-fidelity amplifier and applied to the transducer. The pickup electrode, near the transducer, was connected to a vibration detector whose output was a signal having a frequency f and an amplitude and phase determined by the vibration of the glass blade-transducer system. This signal was fed back into the oscillator locking circuit and the frequency of the oscillator was adjusted to maximize the amplitude of vibration.

The photosensitive element the authors used was an RCA 7326 photomultiplier tube. It was followed by a preamplifier that had two channels: one was tuned to reject the modulation frequency and had a DC output proportional to the light intensity, the other was tuned to amplify the modulation signal and had an AC output that was phase sensitive detected on a commercial instrument and recorded on a strip chart.

When the authors wanted to measure structure on an intense background, the DC output of the preamplifier was used as an error signal for the power supply that polarized the photomultiplier, in order to have a constant DC component, i, of the photomultiplier anode current. In that case, the signal given by the lock in the amplifier was proportional to $(1 \diagdown i_0)$ $(di/d\lambda)$, where i was the photomultiplier anode current and λ the wavelength.

Theory Applied to the Instrument Used (Including Equations)

The authors tried to develop a simple modulation technique which, without affecting any of the optical elements in the monochromator, caused the output wavelength to vary periodically. The modulating device was designed in such a way that an amplitude modulation of the beam occurred only if the derivative of the light intensity with respect to wavelength was nonzero.

A blade of transparent material was mounted on a piece of spring steel tape and vibrations on the system were induced by a transducer attached to a heavy piece of brass. For all the test measurements, the authors used a glass microscope slide ($1 \times 25 \times 76$ mm) as the blade, a piece of 0.01-in. (0.25 mm)-thick steel tape, and a bimorph ceramic transducer ($3.18 \times 10.2 \times 32$ mm). The purpose of the heavy piece of brass was to decouple the transducer from its holder in order to prevent composition modes of vibration to take place. These four pieces were glued together with epoxy and fastened solidly to a fixed part in the monochromator. The resonance frequency of the system decreased when the mass of the glass blade was increased. This was inversely proportional to the square of the free length of the steel tape and was proportional to its thickness. With the above components, the resonance frequency was about 165 Hz when the free length of the steel tape was 2 mm.

The mechanical resonance of the blade-transducer system could be observed optically and electronically. The optical method consisted of (1) making the blade vibrate outside the monochromator and (2) of looking at a point source of light that was reflected on the surface of the blade. In this fashion, the authors could easily find the resonance frequency and at the same time determine the mode of the vibration. The electronic method consisted of placing a pickup electrode near the grounded face of the transducer. A capacitor varying with the blade vibration was thus formed. If a vibration detector was connected to the electrode, one would obtain a signal proportional to the amplitude of vibration at the frequency of vibration.

When a ray of light transversed a plane parallel plate, it underwent a lateral displacement d given by $d = t\sin\varphi(1 - \cos\varphi/n\cos\varphi)$ or $d \propto t\varphi$ ($n - 1/n$) given for small angles, where t was the thickness of the plate, n its index of refraction, φ and φ' the angles that the ray made with the normal of the plate, respectively, outside and inside the plate.

The amplitude of oscillation of the blade was proportional to the excitation voltage applied to the transducer. For the authors' case, a 400-V peak-to-peak voltage at the resonance frequency produced an angular oscillation of ± 0.03 rad. If a ray of light fell at normal incidence on a 1-mm-thick glass blade, such an oscillation would displace it by ± 10 μm. When the oscillating blade was placed just behind the entrance slit of a monochromator that had a dispersion of 100 μ/Å, the lateral displacement of the beam caused a wavelength of ± 0.1 Å.

Because of the change of the optical path length, the glass blade slightly defocused the monochromator and possibly increased the level of the stray light; the instrument therefore lost a little in resolution (the latter effect was more important when the blade was placed behind the exit slit). With a 10-μm slit, for example, the mercury green line given by a General Electric germicidal lamp appeared to have a half-width of 0.35 Å in the absence of the blade; however, when the blade was inserted, this half-width became 0.45 Å.

There was a small amplitude modulation of the light intensity owing to the change in the reflection coefficient with the angle of incidence of the beam on the vibrating blade. If the blade was perfectly perpendicular to the beam, the amplitude modulation signal due to that effect had one component only, at twice the frequency of the blade, and it would not be detected by a phase-sensitive amplifier whose reference signal had a frequency equal to that of the vibrating blade.

The following equations were not a rigorous treatment: their purpose was to show that in certain cases the wavelength modulation eliminated an important portion of the noise that cannot be rejected by the chopper technique.

The light output I of a monochromator was regarded as a superposition of a perfectly constant light intensity I_0 and two randomly fluctuating components N_1 and N_2 were assumed to represent the noise power at a given wavelength. N_1 was the white noise and N_2 was the noise generated by whatever imperfections of the system may prevent the light intensity from being constant, e.g., drift of power supply voltage of the lamp, mechanical vibrations of the system that affect the optical alignment, slow changes in the atmosphere absorption.

N_1 and N_2 were expanded in Fourier series:

$$N_1 = \int_0^\infty \eta_1^c(\omega)\cos\omega t\, d\omega + \int_0^\infty \eta_1(\omega)\sin\omega t\, d\omega$$

$$N_2 = \int_0^\infty \eta_2^c(\omega)\cos\omega t\, d\omega + \int_0^\infty \eta_2^c(\omega)\sin\omega t\, d\omega$$

where η_1 and η_2 had the dimension of noise power per unit frequency range and ω was the angular frequency.

In many cases of interest, e.g., when an incandescent lamp was used, η_2 was large at a very low frequency and decreased rapidly with increasing frequency.

The relative magnitudes of η_1 and η_2 and their wavelength dependence varied with the experimental setup. In many important cases, it was expected that η_2 would have a negligible dependence on the wavelength of the light.

The total intensity output of the monochromator was written as:

$$I = I_0 + \int_0^\infty [\eta^c(\omega)\cos\omega t + \eta^c(\omega)\sin\omega t]d\omega \tag{A-1}$$

where $\eta(\omega)$ was a superposition of $\eta_1(\omega)$ and $\eta_2(\omega)$.

Chopper Technique

The power P received by a noiseless photodetector as a function of time was written as:

$$P = (1/2)I(1 + \cos\omega_c t) \tag{A-2}$$

when the light beam was periodically interrupted at an angular frequency ω_c. Replacing I by Equation A-1 gave

$$P = (1/2)I_0(1 + \cos\omega t) + 1/2 \int_0^\infty [\eta^c(\omega)\cos\omega t + \eta^c(\omega)\sin\omega t]d\omega$$

$$+ 1/2\cos\omega_c t \int_0^\infty [\eta^c(\omega)\cos\omega t + \eta^c(\omega)\sin\omega t]d\omega \tag{A-3}$$

The authors assumed that the response of the phase-sensitive amplifier was constant over a bandwidth $\Delta\omega$ centered at ω_c and zero elsewhere. They further assumed that $\eta(\omega)$ is constant in a range $\Delta\omega$. Under these assumptions, the detected portion P' of the power would be

$$P' = (1/2)I_0\cos\omega_c t + (\Delta\omega/2)\eta^c(\omega_c)\cos\omega_c t + (\Delta/4)\cos\omega t[\eta^c(O) + \eta^c(2\omega_c)] \tag{A-4}$$

The first term was the desired signal, the rest was noise. It was noticed that the low-frequency components of the noise were chopped and, therefore, produced a signal at a frequency $\omega_c \pm \Delta\omega/2$. If $\eta(O) \gg \eta(\omega_c)$, an appreciable amount of noise would be accepted by the amplifier.

Wavelength Modulation

The power M received by the photodetector as a function of time was written as:

$$M = I + (\partial I/\partial\lambda)\Delta\lambda\cos\omega_m t \tag{A-5}$$

where λ was the wavelength, $\Delta\lambda$ the modulation depth, and ω the angular frequency of the modulator. Replacing I in Equation A-5 by Equation A-1 gave

$$M = I_0 + (\partial I_0/\partial\lambda)\Delta\lambda\cos\omega_m t + \int_0^\infty [\eta^c(\omega)\cos\omega t + \eta^c(\omega)\sin\omega t]d\omega$$

$$+ \Delta\lambda\cos\omega_m t \int_0^\infty \left[\frac{\partial\eta^c(\omega)}{\partial\lambda}\cos\omega t + \frac{\partial\eta^c(\omega)}{\partial\lambda}\sin\omega t\right]d\omega \qquad \text{(A-6)}$$

Using the same amplifier considered by the previous case, the authors detected a portion M' of M given by

$$M' = (\partial I_0/\partial\lambda)\Delta\lambda\cos\omega_m t + \Delta\omega\eta^c(\omega_m)\cos\omega_m t$$

$$+ (\Delta\omega/2)\Delta\lambda\cos\omega_m t[\partial\eta^c(O))/\partial\lambda + (\partial\eta^c(2\omega_m))/\partial\lambda] \qquad \text{(A-7)}$$

The first term was the desired signal, the rest was noise.

For a meaningful comparison of P' and M, the authors assumed that ω_c and ω_m were of the same order of magnitude. The fact that the noise contributions to P' and M' were not identical suggested that the wavelength modulation presented significant advantages over the chopped beam technique whenever an important wavelength-independent, low-frequency noise was present $[\eta(O)\gg\eta(\omega_m)]$.

What Was Analyzed

The absorption spectrum of I_2 contained in a commercial sun gun lamp (Sylvania DWY) was measured with the wavelength modulation technique.

Conclusion

The device described allowed the authors to perform derivative spectroscopy, and, therefore, to more easily detect structures on an intense background or weak lines in the vicinity of more intense lines. The authors believed that this technique was particularly useful for Raman effect studies.

The authors believed that the advantages of this device were that it did not appreciably disturb the alignment of the monochromator, it was not very sensitive to external perturbations, it could be tuned to a wide range of frequencies in the audio spectrum, and it could be utilized, with proper modifications, over much of the optical spectrum. It could be removed in a matter of minutes and the monochromator did not require realignment for normal transmission measurements. The electronics associated with the device were commercially available, except for the vibration detector, which could be built without too much difficulty.

FREDERICK R. STAUFFER AND HAJIME SAKAI[108]
DERIVATIVE SPECTROMETRY USING A ROTATING MIRROR TO MODULATE THE SPECTRUM IMAGE

Stauffer and Sakai used a rotating mirror stepped along one diameter to modulate the spectrum image by a discrete amount.

Type of Instrument Used

It was necessary to have an attachment in order to obtain the derivative spectroscopy technique. The exit slit plane of a monochromator (exit slit removed) was imaged onto a final exit slit plane via a circular mirror (referred to as the schizophrenic mirror), stepped along its diameter with each half-plane parallel. Radiation struck this mirror assembly at 45°, and the offset provided a lateral displacement of the spectrum in the final exit slit plane. The rotational motion of this mirror assembly about its shaft provided a switching of the two spectral components in their positions on the final exit slit plane and consequenced an alternative striking of these radiations on the detector.

The displacement of these two spectral components in terms of wave number was given by

$$\delta = (2)^{1/2}d/(dl/d\sigma) \tag{1}$$

where d was the height of the step between the two offset parallel mirrors and $dl/d\sigma$ was the linear dispersion in terms of wave number on the final exit slit plane. Defocusing of the image on the exit slit given by $(2)^{1/2}d$ was negligible. Reasonably good squarewave modulation was effected and the elimination of scattered radiation was mainly due to the homogeneity of scattered radiation with angle. When the two spectral intervals were set adjacent by a proper selection of δ, the difference of these two signals approximated the derivative of the spectrum that would have been obtained from the same spectrometer without the operation of taking the difference.

Theory Applied to the Instrument Used (Including Equations)

Slit Function and Its Transfer Function

Regarding a spectrometer as a linear stationary system, the authors described the observed signal $\psi(\sigma)$ as a convolution between the input signal $\psi(\sigma)$ and the instrumental distortion $g(\sigma)$:

$$\psi(\sigma) = \int \psi(\sigma')g(\sigma - \sigma')d\sigma' \cong \psi(\sigma)*g(\sigma) \tag{2}$$

In spectrometry, the instrumental distortion $g(\sigma)$ was called the slit function. In general, $g(\sigma)$ ceased to become stationary whenever the authors covered a very broad spectral range. However, in normal cases, the authors assumed $g(\sigma)$ to be stationary and utilized by Equation 2. This assumption was workable in practice as a good approximation. With this assumption of stationary $g(\sigma)$, the convolution of Equation 2 was written as a multiplicative form in its Fourier domain A:

$$\psi(z) = \bar{\psi}(z)\tilde{g}(z) \tag{3}$$

where $\tilde{\ }$ designates the Fourier transform. The functions in the z domain were a spatial frequency representation of their Fourier pair. The transform $\tilde{g}(z)$ is called the transfer function.

Transfer Function of the Spectrometer in Regular Motion

When the spectrometer was used in a conventional manner, the slit function that resulted from geometric imaging was a triangular function given by

$$g(\sigma) = 1 - \|\sigma\|/a \qquad \|\sigma\| \leq a$$
$$g(\sigma) = O \qquad \|\sigma\| \geq a \tag{4}$$

where a was the spectral slit width. When the geometric slit width was s, the spectral slit width a is given by

$$a = s/dl/d\sigma \tag{5}$$

The transfer function $\tilde{g}(z)$ corresponding to $g(\sigma)$ in Equation 4 was given by

$$\tilde{g}(z) = [\mathrm{sinc}(az)]^2 \tag{6}$$

with the normalization $g(O) = 1$. The higher spatial frequency content of $\psi(z)$ was suppressed as $[\mathrm{sinc}(az)]^2$. In the σ domain, the normal observation space, this suppression was observed as smearing of the signal by the slit function.

Transfer Function of Derivative Spectrometry

When the authors measured the difference signal of the two spectral components in the derivative mode, they generally obtained a suppression of the lower spatial frequency content in $\check{\omega}\,(z)$. In turn, the authors made an observation with a relative emphasis on the contribution from the higher spatial frequency content in $\check{\omega}(z)$. The transfer function the authors generated in the derivative mode had a structure that made this filtering possible.

The slit function $g(\sigma)$ in this case was given by

$$g(\sigma) = g_1(\sigma) - g_2(\sigma) \tag{7}$$

where $g_1(\sigma)$ and $g_2(\sigma)$ were triangular functions centering at $\sigma = \pm\frac{1}{2}\delta$;

$$g_1(\sigma) = 1 - \|\sigma - \tfrac{1}{2}\delta\|/a \qquad \|\sigma - \tfrac{1}{2}\delta\| \leq a$$
$$= O \qquad \|\sigma - \tfrac{1}{2}\delta\| < a \tag{8}$$

and

$$g(\sigma) = 1 - \|\sigma + \tfrac{1}{2}\delta\|/a \qquad \|\sigma + \tfrac{1}{2}\delta\| \geq a$$
$$= O \qquad \|\sigma + \tfrac{1}{2}\delta\| > a$$

The transfer function $g(z)$ in this case was given by

$$g(z) = 2i\sin\pi\delta z[\mathrm{sinc}(az)]^2 \tag{9}$$

with the normalization $g_1(O) = g_2(O) = 1$. The transfer function was zero at $z = O$, and it had its peak value at $z \neq O$. This was quite contrary to the case for the conventional mode of the spectrometer in which the transfer function had its maxima at $z = O$. With a proper selection of δ and z, $g(z)$ may become a high-pass, low-cutoff filter. This being the case, use of the derivative technique permitted the filtering out of $\check{\omega}(z)$, the unwanted spatial frequency content that existed in the vicinity of $z = O$.

Suppression of Effects from Atmospheric Turbulence

Because the authors recorded the observed signal $\check{\omega}(\sigma)$ as a function of time, its Fourier transform $\check{\omega}(z)$ was in reality registered as a function of temporal frequency. The transfer function $g(z)$ was, therefore, a filter acting in the same domain, the temporal frequency. The authors used (z) given by an earlier equation as a low-frequency cutoff filter to suppress unwanted low frequency fluctuations that were caused by the atmospheric turbulence. In doing so, the temporal frequency characteristics of the transfer function $\tilde{g}(z)$ as a low-cutoff filter were adjusted by changing the wave number drive speed of the spectrometer. If the authors scanned the spectrometer at a speed of $v(d\sigma/dt = \mathrm{cm}^{-1}\mathrm{s}^{-1})$, the actual temporal frequency $f(\mathrm{Hz})$ was given by

$$f = vz \tag{10}$$

where the scale z is expressed as

$$1/\sigma = 1/cm^{-1}$$

Derivation of Spectrum

With small angle approximation, the sine function was given by

$$\sin e\pi\delta z = \pi\delta z \tag{11}$$

The transfer function $\tilde{g}(z)$ of Equation 9, using this approximation, became

$$g(z) = 2i\pi\delta a[\text{sinc}(az)]^2 \tag{12}$$

The observed spectrum $\psi(\sigma)$ was then given by

$$\psi(z) = \delta(i2\pi z)\int_{-\infty}^{\infty}\hat{\omega}(z)[\text{sinc}(az)]^2(\exp - i2\pi\sigma z dz)$$

$$= \delta(d/d\sigma)[\psi(\sigma)*g_0(\sigma)] \tag{13}$$

where $g(\sigma)$ was the triangular slit function of Equation 4. This slit function was the one the authors would have obtained in the regular mode but not in the derivative mode. Because $[\text{sinc}(az)]^2$ had a significant value only in the range $\|z\|<1/a$, the small angle approximation specified by Equation 11 held within this range. The requirement of this small angle approximation to within 5% error was given by

$$\pi\delta z \leq \pi/6 \tag{14}$$

and this range of z fell into the range

$$\|z\| \leq 1/a \tag{15}$$

Thus, the authors obtained the conditions

$$\delta/a < 1/6$$

or

$$\delta < a/6 \tag{16}$$

in order to observe $\psi(\sigma)$ expressed by Equation 13.

Thus, the maximum separation of the two spectral components necessary to obtain a mathematical derivative of the spectrum was one sixth of the spectral slit width within the accuracy stated above.

What Was Analyzed

No analysis was done.

Conclusion

The theoretical analysis made above clearly showed where the advantage lies of the present derivative spectroscopy technique. Use of this technique brought very selective spatial frequency filtering which had the maximum response in the high-frequency range. This filter characteristic was very different from that obtained with regular use of the spectrometer. The latter case had its maximum response at $z = (O)$, while the former had zero response. Because of this suppression of the low-spatial frequency control, the derivative spectroscopy technique offered the advantages over the spectrometer listed in the previous section. The authors appreciated these advantages in part because the detection system that they used was limited by a finite dynamic range.

Special attention was paid to the suppressive feature of the atmospheric turbulence effect by the derivative technique. The low-frequency intensity fluctuations caused by the turbulence were filtered together with the same frequency content of the incoming signal. The atmospheric turbulence effect could not be subtracted out of the signal. Unfortunately, the low-spatial frequency content in $\psi(z)$ played a rather important role in most measurements. The measurement in this case was made with a substantial loss in information content. Compared with this method, the internal modulation technique of Fourier spectroscopy seemed much more powerful in the elimination of the atmospheric turbulence effect, yet it did not destroy the essential information content of the incoming signal.

The technical limitation of derivative spectroscopy was strongly coupled with that of the conventional spectrometry on which it was based. Using Fourier spectroscopy techniques, that measured the spatial frequency spectrum of the signal spectrum, or the interferometer as it was called, one could manufacture the spectrum obtainable from either the regular mode or the present derivative technique by multiplying it by a transfer function having a suitable shape for the final purpose. However, in the situation in which the Fourier spectroscopy technique fails to achieve an advantage over the conventional technique, the authors found a rather powerful use for the derivative technique discussed in this paper as a technique complementary to that of the regular one.

R. N. HAGER, JR. AND R. C. ANDERSON[109]
DEVELOPMENT OF THEORETICAL INTENSITY EXPRESSIONS FOR DERIVATIVE SPECTROMETERS

Type of Instrument Used

This paper offered several generalizations mostly based on only the assumption of a sinusoidally modulated derivative spectrometer instrument. It was first shown, for the mathematical derivative, that the filtering in the transfer function was done by Bessel functions of order equal to that of the derivative. The inverse Fourier transform then gave the output signals as well as an effective slit function for the mathematical derivative spectrometer, in terms not only of Chebyshev polynomials, but also of weighted mean derivative.

Theory Applied to the Instrument Used (Including Equations)

Detector Signal

The authors first considered a direct spectrometer. They let $I_0(\lambda)$ be the true spectral distribution entering the spectrometer, and $I(\lambda)$ the spectral flux at the exit slit when the slit was centered at wavelength λ. They supposed that λ_0 was some specific wavelength about which $I(\lambda)$ could be expanded in the Taylor series

$$I(\lambda) = \sum_{n=0}^{\infty} \frac{1}{n!} - I^{(n)}(\lambda_0)(\lambda - \lambda_0)^n, \; n = 0, 1, 2, \ldots \quad (1)$$

where the notation $I^{(n)}(\lambda_0)$ meant the nth derivative of $I(\lambda)$ with respect to λ evaluated at λ_0.

The authors then supposed that the direct spectrometer was modified in some manner so that the wavelength centered on the exit slit was not constant, but varied sinusoidally in the wavelengths about λ_0. The center wavelength on the exit slit was $\lambda = \lambda_0 + a\sin(\omega t)$, where a was the amplitude of oscillation and ω was the frequency. With such a modification, a $\sin(\omega t)$ was substituted into Equation 1 for $(\lambda - \lambda_0)$ to give

$$I(t) = \sum_{n=0}^{\infty} \frac{1}{n!} - I^n(\lambda_0)a^n\sin^n(\omega t) \quad (2)$$

where $I(t)$ was the measured flux.

It was thus necessary to expand each $\sin^n(\omega t)$ into a finite sum of multiple angle sine or cosine functions. Equation 2 can then be rewritten as

$$I(t) = \sum_{n=0}^{\infty} a^{2n}I^{2n}(\lambda_0)/2^{2n}(n!)^2 + \sum_{n=1}^{\infty} \sin[(2m - 1)\omega t]$$

$$\sum_{n=m}^{\infty} \frac{a^{2n-1}\,I^{2n-1}\,(\lambda_0)(-1)}{2^{2n-2}(n + m - 1)!(n)} + \cos(2m\omega t) \sum_{n=m}^{\infty} \frac{a^{2n}I^{2n}(\lambda_0)(-1)^m}{2^{2n-1}(n + m)!(n - m)!} \tag{3}$$

Equation 3 represented the radiant flux reaching the detector as a function of time and λ_0. The first sum was the DC component of the signal, whereas the other two sums represented the AC components. A standard phase-lock amplifier, tuned to a particular frequency and phase, gave an output voltage proportional to the amplitude of that particular waveform, i.e., proportional to one of the infinite sums for a given value of m, if the detector response was flat.

Transfer Function

The spectral flux out of a spectrometer was written as

$$I(\lambda) = \int_{-\infty}^{\infty} I_0(\lambda')S(\lambda - \lambda')d\lambda'$$

where $S(\lambda - \lambda')$ was the instrument function and $I_0(\lambda')$ was the true incident spectral flux. The Fourier transform of this equation can be written

$$\bar{I}(Z) = \bar{I}_0(\bar{Z})\bar{S}(Z)$$

where the definition of the inverse Fourier transform of $I(\lambda)$ was

$$I(\lambda) = \int_{-\infty}^{\infty} \bar{I}(Z)e^{-2\pi i\lambda z}\,dz$$

It followed that

$$\bar{I}^r(z) = \bar{I}_0(z)\bar{S}(z)(-2\pi iz)^r \tag{4}$$

where $\bar{I}^r(z)$ was the Fourier transform of the rth derivative of $I(\lambda)$ with respect to λ.

Returning to Equation 3, the summations that presented accessible information were denoted as

$$\sum_0(\lambda) = \sum_{n=0}^{\infty} (a^{2n}I^{2n}(\lambda_0)/2^{2n}(n!)^2$$

$$\sum_{2n-1}(\lambda_0) = \sum_{n=m}^{\infty} \left(\frac{a^{2n-1}I^{2n-1}(\lambda_0)(-1)^{m+1}}{2^{2n-2}(n + m - 1)!(n - m)!} \right)$$

$$\sum_{2m}(\lambda_0) = \sum_{n=m}^{\infty} \left(\frac{a^{2n}I^{2n}(\lambda_0)(-1)^m}{2^{2n-1}(n + m)!(n - m)!} \right) \tag{5}$$

The Fourier transforms of each sum were obtained because the only variable functions were derivatives whose transforms were given by Equation 4. Therefore, the transform of $\Sigma_0(\lambda_0)$ was

$$\Sigma_0(z) = \sum_{n=0}^{\infty} \frac{a^{2n}\bar{I}_0(z)\bar{S}(z)(-2\pi i z)^{2n}}{2^{2n}(n!)^2}$$

The summation was recognized as the zero order Bessel function.

Similarly, the remaining two sums were shown to be Bessel functions with integer orders. Therefore, the resulting transformations were

$$\Sigma_0(z) = \bar{I}_0(z)\bar{S}(z)J_0(2\pi a z)$$

$$\Sigma_{n-1}(z) = (-2i)\bar{I}_0(z)\bar{S}(z)J_{2m-1}(2\pi a z)$$

$$\Sigma_{2m}(z) = 2\bar{I}_0(z)\bar{S}(z)J_{2m}(2\pi a z) \qquad m = 1, 2, \ldots \qquad (6)$$

Equation 6 was in the form

$$\Sigma(z) = \bar{I}_0(z)\psi(z)$$

What Was Analyzed

No analysis was made.

Conclusion

The discussion presented is applicable to most of the previously reported derivative spectrometers that utilize a sinusoidally modulated wavelength positioner.

D. T. WILLIAMS AND R. N. HAGER, JR.[110]
SECOND DERIVATIVE SPECTROMETER WITH OSCILLATING EXIT SLITS

Williams and Hager employed an oscillating exit slit to obtain second derivative absorption spectra.

Type of Instrument Used

A small portable second derivative spectrometer, the d^2 spectrometer, was used. This instrument was to be used in the field for the measurement of very small concentrations of certain gases, some of which were air pollutants. When used for studies of air pollution, the d^2 spectrometer was a system made up essentially of three packages. The spectrometer itself was a small Jarrel-Ash quarter-meter grating instrument, Model 82-405. It was mounted on a board that also held the light source with power supply and the absorption cell. The cell was placed between the fixed exit slit of the spectrometer and the photomultiplier tube. The wobbler carrying the oscillating slit was located behind the light source and replaced the fixed entrance slit of the minichromator. The second package of the d^2 system was the d^2 box containing the transistorized tuned phase lock-in amplifier in one end. The third package was the recorder, a Honeywell Electronik 19, modified to produce a signal that was the ratio of the rectified AC voltage to the average DC voltage, both outputs from the d^2 box. Behind the two instruments was the HEW cell, developed for use in detecting air pollutants in the field.

Theory Applied to the Instrument Used (Including Equations)

The absorption of light by gas of concentration c in a light path of length L was given by the relation

$$I(\lambda) = I_0(\lambda)e^{cL\mu(\lambda)} \tag{1}$$

where I was the remaining intensity of the incident intensity I_0 after absorption, the parameter μ was the absorption coefficient and λ was the wavelength.

Modern spectroscopy often utilized a scanning process in measuring I and I_0 so that these two magnitudes were not always measured simultaneously. It therefore followed that if one wished to obtain the concentration c via Equation 1, the limit of sensitivity in the determination was set by the constancy of I_0 as a function of time during the period in which I_0 and I are sequentially determined. If that period is shorter, the drifts in intensity of low frequency are less significant as limiting the precision of determination of c.

One way of obtaining data on I in comparison with I_0 was the use of an oscillating slit in the spectrometer. Suppose such a slit has a position x, which is a harmonic function of time,

$$x = x_0 \sin \omega t \tag{2}$$

The wavelength corresponding to the center of the slit was designated as a function of time by

$$\lambda - \lambda_c = \Delta\lambda \sin \omega t \tag{3}$$

where λ was the wavelength referred to and λ_c was that of the center of the oscillation. The amplitude $\Delta\lambda$ is related to that x_0 of the motion by

$$\Delta\lambda = x_0 d\lambda/dx \tag{4}$$

Where the spectrum being observed was continuous, the intensity of the light beam was expressed in the terms of a power series about the wavelength λ_c

$$I(\lambda) = I(\lambda_c) + (\partial I/\partial\lambda)_c(\lambda - \lambda_c) + (\partial^2 I/\partial\lambda^2)_c(\lambda - \lambda_c)^2/2 + \ldots$$

$$= I_c + (\partial I/\partial\lambda)_c\Delta\lambda \sin \omega t + 1/2(\partial^2 I/\partial\lambda^2)_c\Delta\lambda^2 \sin^2\omega t + \ldots$$

$$+ 1/m!(\partial_m I/\partial\lambda^m)(\Delta\lambda)^m\sin^m\omega t + \ldots \tag{5}$$

The 2*n*th power of the sine function was written

$$\sin^{2n}(\omega t) = \frac{1}{2^{2n}} \sum_{k=0}^{n-1} \left[(-1)^{n-k} 2\binom{2n}{k}\right]$$

$$X\cos2(n - k)\omega t + \binom{2n}{n} \tag{6}$$

and the $(2n - 1)$st power of the sine was written

$$\sin^{2n-1}(\omega t) = \frac{1}{2^{2n-2}} \sum_{k=0}^{n-1} (-1)^{-n-k-1} \binom{2n - 1}{k}$$

$$X\sin(2n - 2k - 1)\omega t \tag{7}$$

The symbol *(p/q)* represented the binomial coefficient:

$$(p/q) = p!/q!(p - q)! \tag{8}$$

Via the use of Equations 6 to 8, Equation 5 was written in terms of the Fourier series:

$$I(\lambda) = \sum_{n=0}^{\infty} \frac{(\Delta\lambda)^{2n}I_c^{2n}}{2^{2n}(n!)^2} + \sum_{m=1}^{\infty} \cos2m\omega t \sum \left(\frac{(-1)^m(\Delta\lambda)^{2n}I^{6n}}{2^{2n-1}(n - m)!(n + m)!}\right.$$

$$+ \sin(2m - 1)\omega t \sum_{n=m}^{\infty} \frac{(-1)^{m-1}(\Delta\lambda)^{2n-1}I_c^{2n-1}(n + m)}{2^{2n-1}(n - m)!(n + m)!2} \tag{9}$$

where I_c was the *k*th order derivative of *I* at the wavelength λ_c.

A suitably tuned amplifier was used to select from the intensity $I(\lambda)$ only those portions having a frequency 2ω, which was twice the wobbler frequency and the mean signal. The amplitudes of these two signals were written as:

$$I_{\text{ave}} = I_c + (\Delta\lambda)^2/4I_c^2 + [(\Delta\lambda)^4 I_c/64] + \ldots \tag{10}$$

and

$$I_{2\omega} = -[(\Delta\lambda)^2/4]I^2 - [(\Delta\lambda)^4I/48] - [(\Delta N^6 I^6/1536)]\ldots \tag{11}$$

Where the usual absorption lines are due to molecular absorbing gases, these two signals are reduced to the respective first terms.

Equations 10 and 11 were written in terms of spectral intensity and its first derivatives. In practice, the only information relating to such intensity was obtained via the use of a spectrometer. The output of this instrument was related to the input by:

$$I(\lambda) = \int_{-\infty}^{+\infty} I_1(\lambda')S(\lambda - \lambda')d\lambda'$$

Here, the output was $I(\lambda)$, the input was $I_1(\lambda')$, the slit function was $S(\lambda - \lambda\rho)$, and δ was the slit width.

The Second Derivative as an Observable

The authors considered

$$I = J(t)f(\lambda)$$

and wrote

$$f(\lambda) = f_0 e^{-\mu cL}$$

and

$$\partial I/\partial\lambda = Jc^{-\mu cL}(df_0^2/d\lambda - f_0 cLd\mu/d\lambda)\partial^2 I/\partial\lambda^2$$

$$= Jc^{-\mu cL}[d^2 f_0/d\lambda^2 - f_0 cLd^2\mu/d\lambda^2 - 2cL(df_0/d\lambda)(d\mu/d\lambda)$$

$$= f_0 c^2 L^2 (d\mu/d\lambda)^2] \tag{12}$$

Then the authors wrote:

$$\partial^2 I/I\partial\lambda^2 = \frac{1}{f_0}\left[\frac{d^2 f_0}{d\lambda^2} - cL\left(\frac{f_0 d^2\mu}{d\lambda^2} + \frac{2df_0 d\mu}{d\lambda d\lambda}\right) + c^2 L^2 f_0\left(\frac{d\mu}{d\lambda}\right)^2\right] \tag{13}$$

If a scan was made through the region in which a spectral absorption line exists but where the slope $df_0/d\lambda$ was constant such that $d^2 f_0/d\lambda^2 = O$ and if, furthermore, the effects of higher order derivatives were assumed to be negligible, then the resulting scan would provide a signal proportional to cL, if the product $(d\mu/d\lambda)cL$ was small enough that the term in $(cL)^2$ could be neglected compared to the second term at the right of

Equation 13. If $df/d\lambda$ were different from zero, but if $d^2f_0/d\lambda^2$ were still zero, the minimum on one side of the maximum would be larger than that on the other due to the term $2cl(df_0/d\lambda)(d\mu/d\lambda)$.

The authors concluded that a reasonable observable for use in measuring small values of cL would be the difference between the maximum and the mean depth of the two minima in the record of the second derivative signal. This number would depend linearly on cL and it would be independent of any slope $df_0/d\lambda$ of the light source spectrum.

What Was Analyzed

Ammonia in blood and ammonia in exhaled breath was analyzed.

Conclusion

The d^2 spectrometer as described was seen to have some special convenience in that it permitted an increase in the amplification of spectrometric data due to suppression of noise emanating from source intensity fluctuations. The presently explored applications indicated that a substantial increase in available sensitivity was obtained for measurements of small concentrations of pollutant gases when compared to direct absorption observations.

K. L. SHAKLEE AND J. E. ROWE[111]
DOUBLE-BEAM WAVELENGTH MODULATION SPECTROMETER

Shaklee and Rowe used a fused silica refractor plate to modulate the reflectance spectra of several solid compounds.

Type of Instrument Used

A double-beam wavelength modulation spectrometer was used. The system utilized two detectors. Wavelength-modulated light emerging from the monochromator was divided into two beams by a neutral density beam splitter (an uncoated fused silica defractor plate was suitable over most of the spectral range covered by the monochromator). The reflected light was focused on a reference detector while the transmitted light was reflected from the sample to a second detector (referred to as the sample detector).

Theory Applied to the Instrument Used (Including Equations)

The authors considered a simple reflectivity experiment in which the reflected light intensity I_R was given by

$$I_R = I_0 R \tag{1}$$

where I was the incident light intensity and R was the sample reflectivity. The normalized derivative of I with respect to the wavelength λ,

$$\frac{I}{I}\frac{dI_R}{d\lambda} = \frac{I}{R}\frac{dR}{d\lambda} + \frac{1}{I_0}\frac{dI_0}{d\lambda} \tag{2}$$

was made up of two different terms. The first term was the normalized derivative of the sample reflectivity and was the quantity the authors wished to determine. The second term was the normalized derivative of the incident light intensity and as such made an undesirable contribution to the wavelength modulation signal.

In the proposed instrument, if the light was wavelength modulated at a frequency Ω_0, then the AC signal (at Ω_0) from the sample photomultiplier detector was proportional to

$$A!(\Omega_0) = [R_S(1 - R_b)dI_0/d\lambda + I_0(1 - R_b)dR_S/d\lambda]\Delta\lambda \tag{3}$$

where R_b, R_S, I_0 and $\Delta\lambda$ were the beam splitter reflectivity, incident intensity, and wavelength modulation amplitude, respectively. In Equation 3 the authors assumed that R_b was independent of the wavelength and that $\Delta\lambda$ was small enough so that only the first derivative terms were important. The DC signal from this detector (with negligible dark current) was proportional to

$$A!(O) = (1 - R_b)R_S I_0 \tag{4}$$

Hence, if the authors normalized the AC signal by holding the DC signal constant with a potentiometric servo system, then the signal at frequency Ω_0 became proportional to

$$A_S(\Omega_0) = (I_0^{-1}dI_0/d\lambda + R_S^{-1} dR_S/d\lambda)\Delta\lambda \tag{5}$$

From similar considerations applied to the reference photomultiplier, it was apparent that the normalized signal from this detector at frequency Ω_0 was proportional to

$$A_R(\Omega_0) = (I_0^{-1}dI_0/d\lambda)\Delta\lambda \tag{6}$$

but the proportionality constant was not necessarily the same for both detectors.

Rather than trying to match the two servo systems to achieve the same normalization of both photomultipliers, the authors adopted the following procedure. The sample detector was normalized to a DC signal of 20 mV. A wavelength-independent modulation of the incident intensity ($\sim$10% at 500 Hz) was added at a frequency Ω_1 different than the wavelength modulation frequency Ω_0. This was accomplished by a neutral density chopper placed before the entrance slit of the monochromator. The signals of the two detectors at frequency Ω_1 were $\sim$10% of their previous respective DC signals. Hence, the normalization of the two detectors would be matched if the two signals at Ω_1 were equal. The authors used a different amplifier (Analog Devices 105B) to obtain the difference of the two signals. The difference signal at frequency Ω_1 was measured with a lock-in amplifier that drove a servo system to adjust the gain of the reference photomultiplier to null, the difference signal at Ω_1. A second lock-in amplifier was used to measure the component of the difference signal at frequency Ω_0 which (according to Equations 5 and 6) was now proportional to $R_s^1 dR_s/d$. The output of the second lock-in amplifier was plotted continuously with a chart recorder. The servo systems used were modified potentiometric recorders that controlled the accelerating potential of the photomultipliers.

What Was Analyzed

Gallium phosphide and indium phosphide were analyzed.

Conclusion

The authors showed that in order to realize the full potential of the wavelength modulation technique, care must be taken to eliminate the contributions of the incident wavelength signal. This was accomplished by using a continuously compensated double-beam detection system. Such a system was sufficiently sensitive for successful measure of the derivative of the reflectivity of solids and to resolve structures not observable in conventional reflectivity measurements. The apparatus has also been used to measure wavelength-modulated spectra. In both types of experiments, the wavelength-modulation spectra gave more sensitive measurements of the line shapes than conventional methods. Hence, this technique was useful for determining the types of absorption processes involved, as well as for studying the effects of thermal broadening.

W. SNELLMAN, T. C. RAINES, K. W. YEE, H. D. COOK, AND O. MENIS[112]

MODULATION OF THE EMISSION SPECTRA OF ELEMENTS

Snellman et al. modulated the emission spectra of elements in a flame using a quartz reflector plate and used the second derivative mode to detect elements in a mixture.

Type of Instrument Used

The instrument used in this study consisted of a 0.75-m Czerny-Turner mount monochromator, $f/6.5$ aperture, curved slits, exchangeable gratings blazed for 300 and 7500 Å with 1180 grooves per nanometer; an oxygen-hydrogen flame with a total consumption burner and an air-acetylene flame with a premixed nebulizer burner; gas flow meters and pressure regulators; an end-on multiplier phototube, EMI 9558AQ, with a 44-mm diameter photocathode and S-20 spectral response; a 0- to 2100-V multiplier phototube power supply, 0 to 30 mA, 0.001% regulation (line and load), $\pm 0.25\%$ accuracy; a phase-sensitive lock-in amplifier with full-scale sensitivity ranges of 100 nV to 500 mV, and a 10-mV strip chart recorder with a 0.2-s full-scale response time.

Theory Applied to the Instrument Used

The instrument provided a technique that permitted the measurement of weak line spectra nested in or on a broad band of continuum and resulted in an improvement in detection limits. In addition, microliter samples were conveniently taken for analysis. The newly designed optical system permitted a rapid repetitive scan of a narrow wavelength region. This scan was synchronized with the AC amplifier. With the AC amplifier tuned to the same frequency as the vibrating quartz plate, the signal produced was the first derivative of the emission intensity with respect to wavelength. The second derivative was obtained at twice the frequency. This mode of operation permitted the measurement of weak line spectra without the interference from background radiation or broad band spectra. When operating at high frequencies (145 Hz), the apparent signal-to-noise ratio was improved by eliminating the low flicker noise of the flame.

What Was Analyzed

This technique was applied to the determination of lithium in 50 μl of solution containing high concentrations of sodium, and to the determination of barium in a calcium matrix.

Conclusion

A flame emission spectrometer using a rapid repetitive scan of a narrow wavelength region has been developed. Using this method of wavelength scanning, the second derivative of the output intensity was measured. The use of this approach to minimize spectral interferences in matrices and the use of microsamples greatly enhanced the potential of flame emission spectrometry and minimized the need for a monochromator of high resolving power. A plate, made to vibrate at 145 Hz, was mounted behind the entrance slit of the monochromator. The AC amplifier was synchronized with the oscillations of the quartz plate. When the amplifier was tuned to twice the frequency of vibration, the second derivative of the spectrum was obtained. This permitted the measurement of weak line spectra nested in or on a broad band or continuum. It was demonstrated that spectral interference due to CaOH bands or a continuum or both are minimized in the measurement of barium. The elimination of interference from bands and flame structure led to an improvement in the detection limits of alkali and alkaline earth elements in the presence of many matrix ions. An analysis was performed with 50 μl of solution, which made it applicable to biochemical and air pollution studies.

A. M. GILLESPIE, JR. AND J. A. HOWELL[113]
DERIVATIVE SPECTROFLUOROMETER USING OPERATIONAL AMPLIFIERS

Gillespie and Howell connected the output of a Perkin Elmer MPF 43A spectrofluorometer to the input of a differentiator circuit constructed from commercially available operational amplifiers.

Type of Instrument Used

The instrument used for obtaining derivative luminescence spectra was a Perkin Elmer MPF 43A spectroflourometer whose output was connected to the input of a differentiator circuit constructed from commercially available operational amplifiers. The operational amplifier circuit is shown in Section I, Figure 1. This circuit was constructed from an Analog Devices Model 194, operational amplifier manifold and Model 106B operational amplifiers.

Theory Applied to the Instrument Used

No discussion of this aspect was made.

What Was Analyzed

The zeroth, first, and second derivative spectra of the following drugs were recorded in various solvents: chlorpromazine hydrochloride, promazine hydrochloride, reserpine, and rescinnamine. Intensity and wavelength dependence as a function of wavelength scanning speed for the second derivatives of the excitation spectra of these compounds were studied. Problems associated with turbid solutions were discussed. The application of derivative techniques to the quantitation of samples containing components with overlapping bands was presented.

Conclusion

The derivative luminescence technique appears to be applicable to a variety of analytical problems and samples such as cross-contamination of drugs, medicated feeds, residues in animal tissues, mixtures of mycotoxins, and other mixtures that likely produce overlapping fluorescence bands. Because the derivative technique measures the rate of change of intensity with respect to the change in wavelength, overlapping bands can be used, one band having minimal influence over the other. For some drugs, the derivative fluorescence technique may help to avoid some of the limitations imposed upon existing methods such as the necessity for the time-consuming separations in which interfering mixtures are encountered.

G. L. GREEN AND T. C. O'HAVER[114]
DERIVATIVE LUMINESCENCE SPECTROMETRY

Most derivative techniques have involved UV and visible spectrometry. However, there have been applications in IR spectrometry, flame emission, and flame absorption spectrometry. Several different techniques have been used to obtain derivative spectra, including electronic differentiation, numerical differentiation, mechanical tachometers, wavelength modulation with synchronous detection, and dual wavelength spectrometry. The application of derivative techniques to luminescence spectrometry was reported by Green and O'Haver in 1974.

Type of Instrument Used

Comparison of the electronic differentiation and wavelength modulation derivative techniques was performed on an Aminco SPF-125 spectrofluorometer (American Instrument Co. Inc., 8030 Georgia Avenue, Silver Springs, MD), which the authors modified for torque motor grating drive. The authors performed wavelength modulation by adding small oscillatory rotation to the grating, in addition to the normal slow scanning rotation. This particular method of wavelength modulation was used rather than the more conventional refractor plate modulator because the authors believed that this avoids the reflective light losses and spectrometer defocusing caused by the refractor plate. The oscillatory rotation was produced by adding an AC waveform to the linear ramp scanning voltage. The AC waveform was a 15-Hz pseudo-triangle wave generated by dividing the 60 Hz power line signal down to 15 Hz and integrating the resulting square wave with a low-pass RC network. The addition of the waveform to the linear ramp was done at the summing point of the torque motor drive operational amplifiers. The wavelength modulation interval was adjusted by a ten-turn potentiometer inserted as a voltage divider between the resistance capacitance, or RC, filter and the driver amplifier. Values of $\Delta\lambda$ from 0 to 100 mm could be selected. A 5-mm $\Delta\lambda$ was used in this work because the authors believed that this was the largest value that would not significantly distort the derivative spectra. The authors used modulation frequencies up to 100 Hz with this system, but the obtainable modulation intervals were restricted at the higher frequencies because of large inertial loading of the torque motor by the relative massive grating and grating mount. In addition, the authors believed that the vibration caused by oscillating the grating at high frequencies and large amplitudes could cause mechanical problems. Green and O'Haver believe that the 15-Hz frequency used in this work was low enough to prevent excessive vibration and to allow the use of modulation intervals as high as 80 nm if desired. Also, because the waveform is synchronous with and a subharmonic of the line frequency, the problem of beat frequencies resulting from stray 60 Hz pickup is eliminated.

A PAR Model 128 lock-in amplifier was used to process the photosignal in the wavelength-modulation mode. The photoanodic signal was pass filtered and applied directly to the lock-in input, the input impedance of the amplifier serving as the anode load resistor. The lock-in output signal was recorded on the y-axis of a Mosely x-y recorder whose x-axis was driven by the scan ramp.

The electronic differentiation system was inserted between the output of the SPF-125 and a chart recorder. The circuit consisted of a low-pass filter amplifier and two stages of frequency-limited differentiation. The electronic differentiator was constructed on an Analog Devices operational manifold. Green and O'Haver performed both the wavelength modulation and electronic differentiation experiments under the same conditions of scan rate (1.0 nm/s), slit width (1.0 nm), and response time (7 s). The two systems were compared by recording the derivative excitation spectra of anthracene in cyclohexane at 25°C. A 150-W Xe arc lamp excitation source was used. The spectra was uncorrected.

Theory Applied to the Instrument Used

The first part of the work reported here was an experimental comparison of two different methods for obtaining wavelength derivatives: electronic differentiation and wavelength modulation. Wavelength modulation coupled with a lock-in amplifier produced an output signal that was proportional to the derivative of intensity with respect to wavelength. An electronic signal differentiator produced an output signal that was proportional to the derivative with respect to time.

What Was Analyzed

Analysis was done on benzopyrene in methanol, pyrene in isopropanol, and anthracene in isopropanol.

Conclusion

Derivative techniques have been shown to be of use in luminescence spectrometry for both qualitative and quantitative applications.

High-quality derivatives of luminescence spectra were obtained with a simple, low-cost electronic differentiator which was easily added to an existing spectrometer. This situation came about due to the particular characteristics of the noise sources in luminescence measurements and to the belief that under most conditions the signal-to-noise ratios of recorded luminescence spectra were very high.

Section IV

Modern Derivative Spectrofluorometers and Spectrophotometers

APPLIED BIOSYSTEMS SPECTROPHOTOMETER WITH 1000S DIODE ARRAY DETECTOR

Applied Biosystems, Ramsey Analytical Division (formerly ABI Analytical, Kratos HPLC Group, 850 Lincoln Center Drive, Foster City, CA 94404) uses a diode array detector for rapid scanning rather than the traditional variable wavelength detector for high sensitivity/low noise applications. The diode array detector performs spectral scans with usable sensitivity of 0.002 AUFS and $<2 \times 10^{-5}$ AU noise under flowing conditions. The manufacturers combined low-noise digital electronics with advanced diode array technology. When the built-in absorbance derivative software of the 1000 is used, changes in the slope for each peak can be viewed. This can help reveal unresolved peaks.

AVIV MODEL 14DS UV-VIS-IR SPECTROPHOTOMETER

Aviv Associates (810 Towbin Avenue, Lakewood, NJ 08701) manufactures a digitally controlled, ratio double-beam recording spectrophotometer. The computer is IBM-PC compatible and uses the MS-DOS operating system. The data system will give any derivative up to the order of the smoothing polynomial. The data system includes an 80386, 20-MHz computer with 1 meg RAM, $3^{1}/_{2}$ and $5^{1}/_{4}$ in. floppy disks, an 80387 mathematical coprocessor, a VGA color monitor, an XT/AT keyboard, and an Okidata 3320 printer.

Besides the derivative function, data manipulation includes sum, difference, product, or ratio of two data sets. It is also possible to add, multiply, or divide data set x- or y-axis values by a constant. In addition, the reciprocal, log, or analog of data set x- or y-axis values can be obtained. Other functions include general fourth order polynomial transform of a data set, interconvert data set between absorbance and percentage of transmission, calculate mean, and linear regression coefficient for any data subset. The computer can smooth by least-squares polynomial fit up to the tenth order. Other data manipulation is possible.

Other features include auto switching between UV, VIS, and IR, optical alignment of the Cary 14 monochromator, new lamp power supply, and slit control to maintain user-selected bandwidth.

BECKMAN DU SERIES 70 SPECTROPHOTOMETER

Beckman Instruments (2500 Harbor Boulevard, Box 3100, Fullerton, CA 92634–3100) fashioned a DU series 70 spectrophotometer. The instrument scans any sample at speeds up to 2400 nm/min. After this, the operator has the choice of ten calculation options (including peak pick, derivatives, and smoothing without delays associated in changing screens). As a result, the data are manipulated as quickly as pressing a key. The wavelength scanning analysis includes first to fourth derivative, logA calculations, real time or recalled graphic display of data, spectral storage for 40 typical scans, background scan stored for repeat analysis, net absorbance calculations using two or three point overlay presentations of multiple functions of spectrum, overlay of up to six spectra, scan trace to access stored data points, selected interpolation of graphic data, option of grid on graphic display, and capability to operate with automatic samplers.

Other features include automatic powerup diagnostics, self-calibration, rail mount with single cell holder, automatic source change, prealigned and easily changed sources, wake-up timer, interval timer, and cycle counter, event alarm, real-time clock, three communications ports (for choice of x-y and/or dot matrix plotters, Peltier controller, and data I/O interface), keyboard with choice of audible beep and high resolution, and high contrast CRT screen.

CE 6700 HIGH-SPEED REFERENCE UV SPECTROPHOTOMETER

Buck Scientific, Inc. (58 Fort Point Street, East Norfolk, CT 06855) created the CE 6700 High-Speed Reference UV Spectrophotometer. One of the features of this instrument is that it is capable of producing first to sixth derivative spectra. The instrument is quite sophisticated and represents a new and very advanced standard of design. All software is standard, coupled with instantaneous mode selection to enable analytical applications to be selected or changed with extreme rapidity.

The instrument has other salient points, including double beam, fast plotting, storage of 60 methods with data, storage of 100 spectra, storage of 100 time plots or kinetic plots, and security protection of all stored methods. Concentration curve and line fittings are provided with the ability to delete or add standards and automatically refit a new curve or line.

The analysis mode contains all the normal quantitative techniques and includes multicomponent analysis as a standard. Also provided is the facility for automatically computing band heights and areas with the application of a correction for background absorption. Curve fitting to the corrected bands is also provided and includes editing facilities.

In addition, the spectrophotometer performs reaction kinetics with or without plots, concentration curve fitting and editing, quantifications and background correction of bands, curve fitting to quantified bands, stripping adding and differencing of spectra, spectral manipulations and reprocessing, autoadaptive scanning for optimum measuring system, turbid sample optical measuring system, fluorescence, reflectance, and gel scanning.

GILFORD RESPONSE UV-VIS SCANNING SPECTROPHOTOMETER

Ciba-Corning Diagnostics Corp. (Gilford Systems, 132 Artino Street, Oberlin, OH 44074) has produced a UV-VIS Scanning Spectrophotometer with a single cell holder. The RESPONSE utilizes temporal dual- and single-beam operating modes. A CRT provides program parameter selection, presentation of data in graphic and tabular formats, and data manipulation. Standard programs include wavelength scanning, multi-wavelength analysis, standard curve, and read sample.

The computing system incorporates a 16-bit microprocessor with over 500 K of processing memory. The computing system allows manipulation of collected data in many different formats including absorbance, percentage of transmittance, first and second derivative, log absorbance, concentration, and ratio. It allows data storage up to 10,000 data points. It performs automatic wavelength and absorbance calibration, along with diagnostic checks of system functions. Routines are available for graphic cursor, automatic peak identification of up to 12 peaks, autopeak integration, baseline corrections, linear and nonlinear curve fitting for kinetic measurements, linear standard curves, and graphic overlays. The computer automatically selects bandwidths, filters, and lamps as well as recognizes accessories placed into the sample compartment. It incorporates a standard RS-232C serial interface for external computer communication.

The interactive operator console has 16 keys along the side of the CRT. The high-resolution CRT displays graphic and tabular data, programs, and allows for system diagnostics. The sealed touchtone keypad provides an audible signal to user entries.

The optics include a 1200 line per millimeter, holographic grating in Littrow configuration, automatic selection of slits, wavelength range from 200 to 750 nm with a stepper motor-driven wavelength drive, four scan speeds, and a quartz halogen and deuterium lamp with automatic lamp changeover when scanning from visible to UV.

MODEL 260 FIBEROPTIC SPECTROPHOTOMETER

Guilded Ware, Inc. (5190 Golden Foothill Parkway, El Dorado Hills, CA 95630) fashioned a fiberoptic spectrophotometer having innovative software that takes spectral derivatives many times. The spectrophotometer is of the scanning variety with a direct drive concave holographic grating control that includes dual detector housing, dual filter holder, one set of five slits, internal 1500-h tungsten halogen lamp and housing, two extra lamps, high-resolution photometric processing board, temperature-compensated high-resolution grating drive, multiplexer control interface, monochromator purge line, 2 MPA part adapters, 2-m test cable with SMA connectors, interface cable to control unit, interface card for control unit, and wavelength calibration filter.

The unit also includes the general Scanner absorbance measurement program, with the examiner and estimator and Pointer manipulation and calibration programs. The menu-driven Scanner software is provided with the Model 260 for control and data handling. In the data collection mode, the user can collect and display a spectrum and then apply a number of mathematical modifications. Options are available to smooth data, average data, find and determine peak areas, display absorbance/transmittances at selected wavelengths, overlay spectra, and insert screen text.

In addition to the Scanner, three other programs are provided with the Model 260. The Estimator program does simple linear second and third order calibrations and predictions for up to ten constituents when the wavelengths of specific absorptions are known. The Examiner allows the user to manipulate a single spectrum and sets of spectra independent of instrument operation. It has the data manipulation features of the Scanner, but it also includes subtraction and multiplication of whole spectra. The Pointer is capable of making repeated measurements at a single wavelength for studies such as chemical kinetics.

HP 8452A UV/VISIBLE SPECTROPHOTOMETER

Hewlett-Packard Co. (39550 Orchard Hill Place, Novi, MI 48376–8024) has fabricated a relatively low-priced spectrophotometer called the HP 8452A. It is a diode array UV/visible spectrophotometer that offers simple operation for the routine laboratory, plus high-performance features for general purpose method development and research.

HP 89531A software provides system control, data processing, automation, and reporting functions. The software is supported on many different MS-DOS PCs, including the full range of HP Vectras, and the IBM-PC, PC-XT, and PC-AT.

Using the calculation mode, derivatives are used to enhance spectral differences. Other functions include converting spectra to and from absorbance and transmittance modes, detecting impurities and spectral differences by subtracting or dividing spectra, investigating chemical interactions in mixtures by adding standard spectra together and comparing the theoretical mixture spectrum with a real mixture spectrum.

Data can be stored on disc and recalled using descriptive file names. The handling of large quantities of data can be simplified by organizing files into directories and subdirectories.

Hard copy reports of spectral data and results can be produced in appropriate alphanumeric or graphic formats on several types of printers and plotters. Reports automatically contain relevant identification such as sample name, measurement parameters, and date.

MODEL F-4010 FLUORESCENCE SPECTROPHOTOMETER

Hitachi Instruments, Inc. (Miry Brook Road, Danbury, CT 06810) has created a fluorescence spectrophotometer interfaced to a PC. The Model F-4010 is a fully computerized fluorescence spectrophotometer that provides data processing in a simplified, easy-to-use fashion.

The derivative function minimizes the effect of background, enhances spectral definition of peaks otherwise hidden in the major bands, and improves the analytical accuracy by revealing shoulders on spectra as clear independent peaks.

The instrument can be used to produce synchronous fluorescence (SFS). Significant improvement in the selectivity and sensitivity of fluorescence measurements can be obtained from the use of SFS techniques. In the SFS method, the excitation and emission monochromator wavelengths are scanned synchronously while maintaining a constant separation in wavelength between the two monochromators.

Fully corrected excitation and emission spectra is a standard feature of the F-4010. Correction factors based on the rhodamine quantum counter in the 200 to 600 nm range are generated and stored.

The Model F-4010 keyboard and soft key interaction with the graphic display provides a simplified operation. The user can set the instrument operating conditions by using the PARAMETER hard key. The F-4010 can store as many as ten different analytical conditions and calculation programs. The storage number and analytical items are displayed on the CRT for method selection. A Pre-Scan function allows the F-4010 to determine automatically the optimum excitation and emission wavelengths for unknown samples.

To perform an excitation or emission wavelength scan, the user must select an appropriate EX or EM wavelength start/stop key on the keyboard. The CRT displays the measurement graphic screen and starts the wavelength scan. During the wavelength scan, the spectrum is displayed on the CRT in real time. After the measurement is completed, the user can access the data processing functions.

PCI PHOTON COUNTING SPECTROFLUOROMETER

ISS, Inc. (309 Windsor Road, Champaign, IL 61820) has made a photon counting spectrofluorometer for high-sensitivity fluorescence detection. The software (ISSPC Spectra Software, ©1990 ISS) includes routines for the manipulation and analysis of spectra. In particular, it calculates the derivative of a spectrum until the nth order, for each n. Other routines include the calculation of spectral moments (the area under a spectrum, the average and the standard deviation) for the characterization of quantum yield and energy transfer of fluorophores and the smoothing of spectra. Another useful feature for the characterization of heterogeneous samples is the ability to display excitation-emission matrices of spectra and the detection of synchronous luminescence spectra.

ISS also manufactures a multifrequency phase fluorometer, the K2, a powerful instrument allowing the characterization of heterogeneous samples through the measurement of the fluorescence lifetime.

JASCO MODEL PP 777 SPECTROFLUOROMETER

JASCO International Co. Ltd. (4-21 Sennin-cho-2-chome, Hachioji, Tokyo 193, Japan) has manufactured a dual-beam spectrofluorometer. The software is capable to generate one to four derivatives; store eight spectra; retain parameters in battery backup memory of up to eight files; automatically print out photometric values and wavelengths; repeat scanning up to 200 times; manipulate data of addition, subtraction, multiplication, and division; correct a spectrum between 220 and 600 nm. Seven kinds of calibration curves were provided for quantitative analysis.

The main unit has a photometric system with dual beams and correction of light source fluctuation by diode feedback. It has a 150-w ozone-free xenon lamp light source. Its optics include monochromators for excitation and emission. The wavelength range was 220 to 750 nm with a wavelength accuracy of ± 5 nm and a spectral bandwidth of 1.5, 3, 5, 10, and 20 nm on both the excitation and emission monochromators. The wavelength scanning speeds vary from 1 to 4000 nm/min. The unit has a 14-in. color CRT that has high resolution.

UVIKON 940 SPECTROPHOTOMETER

The Uvikon 940 spectrophotometer built by Kontron Instruments (Research Instruments International, 5929 Portobello Court, San Diego, CA 92124, the North American importer for Kontron Spectroscopy products) has the ability to perform curve data manipulation. Wavelength spectra curve data can be added, subtracted, multiplied, and divided by another curve and by numbers. First through fourth derivatives of spectra and time drive curves with three selectable smoothing algorithms can be accomplished. Other manipulation includes the ability to overlay a derivative normalized to the same y-axis distance as the zero order data and interconversion of data between absorbance.

Wavelength scanning includes real-time self-optimization of scanning speed based on the steepness of the peak. The instrument also has selectable cutoff filter-change wavelengths.

Optics embraces a keyboard-selected slit changer for spectral bandwidths of 0.3, 0.5, 1, 2, 4, and 6 nm. In addition, the instrument contains two reduced-height slits (2 and 4 nm spectral bandwidth) for use with reduced-height cuvettes.

The hardware used is an 80286 microprocessor (10 MHz). The software includes a tutorial program which was written to give new users a quick overview of the 940s main features and how to implement them. The tutorial program was written using the 940s Macro Programming Language (MPL) in order to demonstrate some of the MPLs versatility. Source code is provided.

Communications comprise a built-in bidirectional RS232C interface. The 940s RS232C interface allows complete control of all instrument functions by an external computer. All acquired and calculated data can be transferred, including temperatures measured by the Uvikon's optional temperature sensors. Data transfer can be instructed to occur either in real time or automatically at the end of data acquisition.

The Uvikon 940s MPL brings to the user the ability to save time during complex analysis, and it provides the option of customized screen prompts and print outs. Users can create "User Applications" using the MPLs ready-to-use subprograms and functions which are implemented with QuickBASIC and Turbo and Pascal languages. Factory-written example user applications are provided, such as "940 Tutorial" as well as three-wavelength correction to determine proper peak height when scattering is present. Also provided are "include" files and library files.

The output includes a P-900 four-color printer/plotter. The P-900 uses a thermal transfer head for fast, quiet printing of text. For graphics, the P-900 uses four separate ink pens in the same manner as a conventional plotter. The Uvikon 940 also supports the Epson LX series of printers.

OPTOMETRICS RECORDING SPECTROPHOTOMETERS (RS SERIES)

Optometrics USA, Inc. (P.O. Box 699, Nemco Way, Ayer, MA 01432-0699) single-beam Recording Spectrophotometers (RS series) are comprised of interchangeable modular units that enable the user to optimize instrumental parameters for a specific application while retaining system flexibility.

RS systems are designed around the Mini-Chrom monochromator. The user is able to customize the recording spectrophotometer for a single application or a variety of experiments. Ordinate sensitivity can be displayed in %T, Abs, or first and second derivatives.

All RS systems are compatible with PC, XT, and AT types of personal computers. The operating software supplied is menu driven and allows for manipulation of the sample data.

The instrument has several other features. Peaks can be automatically identified and listed with the corresponding Abs or %T value. Detectors and monochromators can be changed to enhance overall performance in a selected spectral region. A sample compartment with in-line and 90° detector ports increase measurement options. The sample stage is capable of holding transmissive and opaque solids as well as solutions.

The instrument contains several menus including the System menu, which can be used to address subroutines; the Parameters menu, which allows selection of wavelength scan or time base mode; the Run menu, which allows acquisition and storage of 1000 data points from both reference and sample scans; the Display menu, which lists the ways in which the previously acquired data can be manipulated and displayed; and the File menu, which allows the run data of any sample to be stored in a directory under a specified file name.

ULTRASPEC PLUS SPECTROPHOTOMETER

ULTRASPEC PLUS UV/Visible integrated spectrophotometer is built by Pharmacia LKB Biotechnology (800 Centennial Avenue, Piscataway, NJ 08854). This instrument has many features including calculation and display of first, second, and fourth order derivative and log spectra.

A software package is available for the ULTRASPEC III. This enables full control from a PC and upgrades the spectrophotometer into a flexible scanning instrument with good data manipulation properties. The color software operates in the Microsoft® Windows environment with a mouse-driven user interface. It can be used for enzyme kinetics, wavelength scanning, and standard curve applications. Features include on-line help pages, storage of methods and results to disk for subsequent data manipulation, transfer to spreadsheets, postrun editing or archiving, and the ability to download onto a wide range of printing peripherals.

The wavelength scanning mode has many features. This includes user-defined zoom facility; spectral addition, subtraction, multiplication, and division of spectral data by a user-entered factor; repeat scans for simple kinetics and isobestic studies; analysis of spectra for peaks, troughs, and inflections; and overlay of spectra.

Standard curve applications include multistandard calibration curve generation, calibration curve storage, postrun editing of calibration curves, linear interpolation, linear regression, and automatic control of rapid sampling accessories.

The ULTRASPEC III UV/Vis and Visible spectrophotometers have spill-proof membrane keypads that enable parameters such as wavelength, function operating mode, and factor to be selected by single keystrokes.

The sample compartment of the ULTRASPEC III is fitted as standard with an automatic six-position cell changer for rapid selection via the keyboard.

The large display has continuous readout of wavelength, lamp status, mode measurement, and cell number. Hard copy of annotated results can be obtained automatically using an external printer.

The ULTRASPEC III goes through a microprocessor-controlled calibration procedure at powerup, ensuring accuracy and wavelength reproducibility. Diagnostic parameters, such as lamp life, are obtained directly from the keypad; lamps can be changed by the user.

PU8700 SPECTROPHOTOMETER

The PU8700 Spectrophotometer was created by Phillips Scientific (York Street, Cambridge, UK). Using the PU8730 software allows the user to obtain derivative and log results, as well as spectrum scanning (including absorbance/transmission against wavelength), scan speeds from 25 to 2000 nm/min, automatic peak and valley search, and a variety of hard copy formats. The software also allows rate assays with absorbance/transmission against concentration, a calibration curve with up to ten standards, four curve fitting routines including quadratic, up to triplicate standards, and continuous on-line monitoring mode.

The software can be used for fixed wavelength modes such as absorbance/transmission at discrete wavelengths; one, two, or three lambda calculations; peak area or height calculations; difference, sum, or ratio modes; and up to eight wavelengths read sequentially.

M SERIES SPECTROFLUOROMETER

Photon Technology International Inc. (1 Deerpark Drive, Suite F, South Brunswick, NJ 08852) has manufactured a fluorometer, which, among other features, generates derivatives to any order. Other interrogation and manipulation features include unlimited numbers of datasets that may be plotted and displayed simultaneously (dataset values may be interrogated, edited, and captured for use as imbedded processing constants in the concentration equation); linear logarithmic and concentration scaling for the y-axis; fixed or autoscaled ranges; unlimited multistage zooming; and panning for magnified displays.

Mathematical functions include linear scaling (multipliers and offset); integration with fully adjustable integration window, a dynamic report of total and peak value integrals; linear least-squares fits to user-specified X range, with report of slope, intercept, and correlation coefficient; Savitsky-Golay running polynomial smoothing; reciprocal calculation; log and antilog functions; ratio to concentration mapping, with user-defined concentration limit values and full viscosity correction; and multitrace combinatorics.

Data acquisition features incorporate the following: all channels of raw data may be saved independently of display function; wavelengths may be preset or scanned; photon-counting acquisition times for each data point conform with the user-specified time base interval or the desired integration time in wavelength mode; raw data may be saved manually or automatically in sequential file sets; background values may be subtracted from all data for autofluorescence and dark current corrections; data may be displayed on the screen in real-time with fixed or dynamically autoscaled y-axis ranging; automatic mode may permit all instrument control and data acquisition, processing, and storage operations to be carried out by a user-developed program for automatic repetition of a defined procedure; and data acquisition may be paused/resumed at any time.

SPF-500 C SPECTROFLUOROMETER

SLM Instruments, Inc. (810 West Anthony Drive, Urbana, IL 61801) have produced a unique spectrofluorometer that can provide first through nth order derivatives. Other features include wavelength scanning and timebase, single polarization, storage of data sets, spectral overlays, difference spectra, spectral smoothing, repetitive scanning, single point %T measurements, postacquisition rescaling and replotting, millisecond kinetics, absorbance and %T spectra, prescan for automatic y-axis setting, postacquisition macros, macrocommand programming, storage and recall of setup protocols, variable delay between scans, math on a single spectrum, math between spectra, integration of an area under a curve, normalization of data around a single point, windowing for plot magnification, record information listing, transfer of data to another disk, editing records, change in printer type, and color CRT display.

The SPF-500C performs a variety of operations and calculations. The instrument includes an IBM desktop computer and a comprehensive menu-driven software package. Bidirectional communication between the IBM desktop computer and the instrument makes the computer a control center and an analytical tool.

FLUOROLOG-2 SPECTROFLUOROMETER

The FLUOROLOG-2 Spectrofluorometer is built by SPEX Industries, Inc. (3880 Park Avenue, Edison, NJ 08820). The main part of the FLUOROLOG-2 spectrofluorometer system is the DM3000 spectroscopy computer. Two of the mathematical operations are derivatives and integrals. The IBM PC/XT compatible DM3000-F offers completely automated control and management of luminescence experiments via user-friendly, menu-driven software. The integrated features of the

DM3000-F also include dual high-voltage power supplies, scan controls and photon counting, and direct-current data acquisition modules — all under software control.

The SPEX software packages supplied with each DM3000-F allows the user to use keystroke programming to create method routines specific to certain experiments, and routines can be stored on disk to automate subsequent work with a particular method. A label mode permits custom labeling of spectra to denote the characteristics of a particular experiment. These systems enable a number of people to pursue independent research with the same experiment by permitting the recall of each researcher's experimental parameters at any time. Parameter control functions include spectrometer wavelength positions, scan limits, data averaging, phosphorimeter delay and sample window, autopolarizer rotation, and excitation shutter position.

Mathematical operations cover arithmetic functions, logs and antilogs, minimum and maximum wavelengths, and peak areas.

Data processing algorithms incorporate variable point Savitsky-Golay smoothing, automatic spectral combinations, background subtraction, and radiometric excitation and emission correction.

Plotting routines comprise file overlays for comparison of multiple spectra, vertical and horizontal expansion, autoscaling to display spectra between maximum and minimum intensity, and cursor-location coordinates.

References

1. Prensky, S. D., Op Amp Experiments Manual, E & L Instruments, Inc., Derby, CT, 1974.
2. Griese, A. and French, C., *Appl. Spectrosc.,* 9, 78, 1955.
3. Collier, G. and Singleton, F., *J. Appl. Chem.,* 6, 495, 1956.
4. Bonfiglioli, G. and Brovetto, P., *Appl. Opt.,* 3, 1417, 1964.
5. Staufer, F. and Sakai, H., *Appl. Opt.,* 7, 61, 1968.
6. Hager, R. and Anderson, R., *J. Opt. Soc. Amer.,* 60, 1444, 1970.
7. Martin, E. W., et al., Eds., *Remington's Practice of Pharmacy,* 12th ed., Mack Publishing, Easton, PA, 1961, 1052.
8. Anderson, C. R., *Ways to Health,* Southern Publishing Association, Nashville, TN, 1962, 485.
9. Martin, E. W., et al., Eds., *Remington's Practice of Pharmacy,* 12th ed., Mack Publishing, Easton, PA, 1961, 1057.
10. Ibid., p. 765.
11. Hercules, D., Ed., *Fluorescence and Phosphorescence Analysis,* Interscience, New York, NY, 1967, 88.
12. Williams, R. T. and Bridges, J. W., *J. Clin. Pathol.,* 17, 375, 1964.
13. Miles, C., Fluorescence of Acetylsalicylic Acid and Salicylic Acid, M.Sc. thesis, Wayne State University, Detroit, MI, 1972.
14. Wilson, D., Analytical Luminescence Spectrometry of Some Antihistamines, Ph.D. thesis, Wayne State University, Detroit, MI, 1972, 10.
15. Ibid., p. 45.
16. Ibid., p. 23.
17. Ibid., p. 20.
18. Riddick, J. A. and Toops, E. E., Jr., Eds., *Organic Solvents, Physical Properties and Methods of Purification,* Interscience, New York, 1955, 270.
19. Martin, E. W. et al., Eds., *Remington's Practice of Pharmacy,* 12th ed., Mack Publishing, Easton, PA, 1961, 186.
20. Clark, G. L., Editor, *The Encyclopedia of Spectroscopy,* Reinhold, New York, 1960, 244.
21. Schenk, G. H., *Absorption of Light and Ultraviolet Radiation, Fluorescence and Phosphorescence Emission,* Allyn & Bacon, Boston, 1973, 8.
22. Perkin-Elmer Corp., Derivative Spectroscopy, Perkin-Elmer Application Data Bulletin, 1977, 7.
23. Perkin-Elmer Corp., Second Derivative Spectroscopy, Perkin-Elmer Application Data Bulletin, 1977, 18.

24. Clark, G. L., Ed., *The Encyclopedia of Chemistry*, Reinhold, New York, 1966, 823.
25. Olsen, E. O., *Modern Optical Methods of Analysis*, McGraw-Hill, New York, 1975, 452.
26. Ibid., p. 454.
27. Ibid., p. 455.
28. Ibid., p. 456.
29. Ibid., p. 459.
30. Perkin-Elmer Corp., Perkin-Elmer Brochure Order No. L-574, 1978.
31. Hoover, J. E., Managing Ed., *Remington's Pharmaceutical Sciences*, 14th ed., Mack Publishing, Easton, PA, 19, 835, 1970.
32. Ibid., p. 1574.
33. Ibid., p. 835.
34. Savitskyu, A. and Golay, M., *Anal. Chem.*, 36, 1628, 1964.
35. Poro, T. J., *Anal. Chem.*, 44, 93A, 1972.
36. Green, G. L. and O'Haver, T. C., *Anal. Chem.*, 46, 2191, 1974.
37. Hoover, J. E., Ed., *Remington's Pharmaceutical Sciences*, 13th ed., Mack Publishing, Easton, PA, 19, 1194, 1965.
38. *United States Pharmacopeia*, 20th ed., USP, Rockville, MD, 19, 621, 1980.
39. Ibid., p. 621.
40. Hoover, J. E., Ed., *Remington's Pharmaceutical Sciences*, 14th ed., Mack Publishing, Easton, PA, 19, 1102, 1970.
41. Ibid., p. 1102.
42. *United States Pharmacopeia*, 20th ed., USP, Rockville, MD, 19, 4, 1980.
43. Ibid., p. 274.
44. Osel, A. and Farrar, G., Eds., *The Dispensatory of the United States of America*, 25th ed., Lippincott, Philadelphia, 1955, 1658.
45. Ibid., p. 1655.
46. Ibid., p. 1783.
47. Ibid., p. 1791.
48. Heilbron, I. and Bumbury, H. M., Eds., *Dictionary of Organic Compounds*, Oxford University Press, New York, 1953, 409.
49. Ibid., p. 232.
50. Osel, A. and Farrar, G., Eds., *The Dispensatory of the United States of America*, 25th ed., Lippincott, Philadelphia, 1955, 1122.
51. Ibid., p. 1391.
52. Ibid., p. 1204.
53. Ibid., p. 1205.
54. Ibid., p. 1205.
55. Conway, E. J. and Cooke, R., *Biochem, J.*, 33, 457, 1939.
56. Tobe, B. A., *Can. Med. Assoc. J.*, 84, 767, 1961.
57. Gangoli, S. and Nicholson, T. F., *Clin. Chim. Acta*, 14, 585, 1966.
58. Seligson, D. and Hirahara, K., *J. Lab Clin. Med.*, 49, 952, 1957.
59. Burg, P. V. D. and Mook, H. W., *Clin Chim. Acta*, 8, 162, 1963.
60. Ballot, A. F. K. and Steendijik, C., *Clin. Chim. Acta*, 12, 55, 1965.
61. *Official Methods of Analysis*, 13th ed., Association of Official Analytical Chemists, Washington, DC, 1980, Sect. 33057.

62. Hurteau, M., Mislan, J. P., and Ashley, R. W., Analysis for Gadolinium in Heavy Water by Flame Emission, Rep. No. AECL-4772, Atomic Energy of Canada Limited, Ottawa, 1974; also available as *Anal. Abstr.*, 28(5), 23, 1975.

63. Atsuya, I. and Goto, H., *Anal. Chim. Acta*, 65, 303, 1973.

64. Munshi, K. N. and Dey, A. K., *Chim. Anal.*, 53(8), 539, 1971; also available as *Anal. Abstr.*, 22, 3047, 1972.

65. Vekhande, C. and Munshi, K. N. J., *Indian Chem. Soc.*, 52, 939, 1975; also available as *Anal. Abstr.*, 31(2), 87, 1976.

66. Bard, J. A., *Encyclopedia of Electrochemistry of the Elements,* Vol. 6, Marcel Dekker, New York, 1976, 34.

67. Neims, J. R. and Vogel, R. S., *Appl. Spectrosc.*, 21, 242, 1967.

68. Vdovenko, M. E., *Zav. Lab.*, 32, 785, 1966; also available as *Anal. Abstr.*, 14, 6724, 1967.

69. Budanora, L. M. and Pinaera, S. N., *Zh. Anal. Chim.*, 20, 320, 1965; also available as *Anal. Abstr.*, 13, 5447, 1966.

70. Vdovenko, M. E. and Lisichenok, S. L., *Zav. Lab.*, 33, 1372, 1967; also available as *Anal. Abstr.*, 16, 598, 1969.

71. Snell, F. D. and Ettre, L. S., Eds., *Encyclopedia of Industrial Chemical Analysis,* Interscience, New York, 1973, 481.

72. Ibid., p. 489.

73. Ibid., p. 490.

74. Ibid., p. 470.

75. Walker, C. S., *Nucl. Saf.*, 7, 45, 1965.

76. Connors, K., Amidon, G., and Kennan, L., *Chemical Stability of Pharmaceuticals*, Wiley-Interscience, New York, 1979, 123.

77. Koshy, K. T. and Lach, J. L., *J. Pharm. Sci.*, 50, 113, 1961.

78. Higuchi, T. and Brochman-Hansen, E., *Pharmaceutical Analysis*, Interscience, New York, 1961, 551.

79. Kornay, M. A., Heber, D., and Schnekenburger, J., *Talanta*, 29, 332, 1982.

80. Davis, D. R., Fogg, A. G., Burns, D. T., and Wragg, J. S., *Analyst*, 99, 12, 1974.

81. Kornay, M. A., Wahbi, A. M., and Hewala, I. I., *Arch. Pharm. Chem. Sci. Ed.*, 12, 26, 1984.

82. Abel-Hamid, M., Kornay, M. A., and Bedair, M., *Acta Pharm. Jugosl.*, 34, 183, 1984.

83. Such, V., Travest, J., and Gonzalzo, R., *Anal. Chem.*, 52, 412, 1980.

84. El-Yazbi, F. A. and Barary, M., *Anal. Lett.*, 18(B5), 629, 1985.

85. Kityamura, K. and Majima, R., 55(1), 54, 1983.

86. Kityamura, K., Takagi, M., and Hozumi, K., *Chem. Pharm. Bull.*, 33(4), 1484, 1984.

87. Mazzeo, P., Quaglia, M. G., and Segnalini, F., *J. Pharm. Pharmacal.*, 34, 470, 1982.

88. Fell, A. F. and Davidson, A. G., *J. Pharm. Pharmacol.*, 32(Suppl.), 97, 1980.

89. Osel, A. and Farrar, G., Eds., *The Dispensatory of the United States of America*, 25th ed., J. E. Lippincott, Philadelphia, 1955, 13.

90. *Pharmacopeia of the United States*, 19th ed., Mack Printing, Easton, PA, 1975, 629.

91. *British Pharmacopeia*, University Printing House, Cambridge, MA, 1973, 472.

92. Hashmi, M., *Assay of Vitamins in Pharmaceutical Preparations,* John Wiley & Sons, London, 1973, 49, 188.

93. Stroheckerr, R. and Henning, H. M., *Vitamin Assay Tested Methods,* Verlag Chemie, Weinheim, Germany, 1965, 65, 82, 129.

94. Menick, D., Hochberg, M., Himes, H. W., and Oser, B. L., *J. Biol. Chem.,* 160, 1, 1945.

95. *Official Methods of Analysis,* 10th ed., Association of Official Analytical Chemists, Washington, DC, 1965, Sect. 17001A.

96. Holum, J. R., *Fundamentals of General, Organic and Biological Chemistry,* 2nd ed., John Wiley & Sons, New York, 1982, 522.

97. Walters, S. M., *J. Chromatogr.,* 259, 227, 1983.

98. Walters, S. M., Westerby, B. C., and Gilvydis, D. M., *J. Chromatogr.,* 317, 533, 1984.

99. Zhou, M. and Miles, C. J., *J. Assoc. Off. Anal. Chem.,* 74(3), 546, 1991.

100. Vuik, J., *J. Chromatogr.,* 553, 299, 1991.

101. Pachinger, A., Eisner, E., Begutter, H., and Klus, H., *J. Chromatogr.,* 558, 369, 1991.

102. Giese, A. T. and French, C. S., *Appl. Spectrosc.,* 9, 78, 1955.

103. Collier, G. L. and Singleton, F., *J. Appl. Chem.,* 6, 495, 1956.

104. Bonfiglioli, G. and Brovetto, P., *Appl. Opt.,* 3, 1417, 1964.

105. Aramu, I. and Rucci, A., *Rev. Sci. Instrum.,* 37, 1696, 1966.

106. Bonfiglioli, G., Brovetto, P., Busca, G., Levialdi, S., Palmiere, G., and Wanke, E., *Appl. Opt.,* 6, 447, 1967.

107. Perregaux, A. and Ascarelli, G., *Appl. Opt.,* 7, 2031, 1968.

108. Stauffer, F. R. and Sakai, H., *Appl. Opt.,* 7, 61, 1968.

109. Hager, R. N., Jr. and Anderson, R., *J. Opt. Soc. Am.,* 60, 1444, 1970.

110. Williams, D. T. and Hager, R. N., Jr., *Appl. Opt.,* 9, 1597, 1970.

111. Shaklee, K. L. and Rowe, J. E., *Appl. Opt.,* 9, 627, 1970.

112. Snellman, W., Rains, T. C., Yee, K. W., Cook, H. D., and Menis, O., *Anal. Chem.,* 42, 394, 1970.

113. Gillespie, A. M. and Howell, J. A., *EDRO SARAP Res. Tech. Rep.,* 5, 1, 1980.

114. Green, G. L. and O'Haver, T. C., *Anal. Chem.,* 446, 2191, 1974.

115. Mills, P. A., *J. Assoc. Off. Anal. Chem.,* 51, 29, 1968.

Q&A

Answers to Experiment 1 Questions

1. What are capacitors?

Capacitors are parallel plate devices made up of a positively and a negatively charged metal plate with air between the plates. These are also called condensers.

2. What is the purpose of a capacitor?

A capacitor stores an electrical charge.

3. What is a breadboard?

A breadboard is a circuit design instrument. It may be an instrument package for linear design work or a socket combined with a speaker, breadboarding pins, battery, clips, mounting speakers, and a test switch that can be used to build and test operational amplifier circuits.

4. Describe a voltmeter and its purpose.

A voltmeter is an instrument used to measure voltage. the voltage is applied to a suitable resistor, the current is measured, and Ohm's law is applied. The resistor is built into the voltmeter, and the scale calibrated directly into voltage.

5. List the differences between a dual in-line and a TO case operational amplifier.

The main difference is the shape. For example, a 16-pin dual in-line package has 16 short pins attached to the operational amplifier body while one type of TO case operational amplifier has eight relatively long pins attached to the operational amplifier body.

6. *What are resistors and how is the ohm value of a resistor determined?*

Resistors are devices that impede the flow of current. They may be made of carbon, metals, or wire wound around various substances. The value of resistors is determined from the different colored bands or different color codes painted on the resistors.

7. *In relation to the operational amplifier, what is ground?*

There are two inputs to the differential input stage. One is called the inverting input and is normally shown as $(-)$, the other is called the noninverting input and is normally shown as $(+)$. There is no ground connection for the operational amplifier, and the circuit works best with split values of equal amounts ranging from $+5, -5$ to $+20, -20$. Ground for the circuit is simply a stake driven in the ground that tells us where the halfway point is between the two supply limits.

Answers to Experiment 2 Questions

1. *What are the first and second luminescence derivatives?*

The theory of derivative spectroscopy involves Beer's law:

$$A = \ln I_o/I = abc$$

here A is absorbance, a is the absorption coefficient that varies with wavelength, and c is the concentration of the sample being analyzed. At a given wavelength, the first derivative is

$$dI/d\lambda/I = (IdI_o/I_o d\lambda) - bc(da/d\lambda)$$

and the second derivative is

$$d^2I/d\lambda^2/I = I/I_o d^2I_o/d\lambda^2 + (bc\ da/d\lambda)^2$$
$$- 2/I_o(dI_o/d\lambda)(da/d\lambda)bc - bc(da^2/d\lambda^2)$$

2. What is the purpose of an amplifier?

The principal application of amplifiers is based on their ability to amplify electrical signals.

3. How does an amplifier work?

The current in a vacuum diode can be controlled by including a third electrode, the grid, in the space between the anode and the cathode. Very little power is consumed by the grid in controlling the anode current so that the triode is an effective amplifier. the transistor supplanted vacuum tubes as an amplifier in 1948.

4. How does an operational amplifier work?

The performance of transistor and vacuum tube amplifiers is enhanced in many respects by returning a fraction of the output signal to the input terminals. This process is called feedback. the feedback signal may either augment the input signal or tend to cancel it. The latter is called negative feedback. Improved frequency-response characteristics and reduced waveform distortion are attained with negative feedback. In addition, amplifier performance is much less dependent on changes in tube or transistor parameters caused by aging or temperature effects.

A particular form of negative feedback, known as operational feedback, is used in amplifiers that perform mathematical operations such as addition or integration on an input signal. Commercially available high-performance de-coupled units have come to be called operational amplifiers. Operational amplifiers are widely used in measurement and control applications, as well as in electronic analog computers.

5. Draw a block diagram of an operational amplifier.

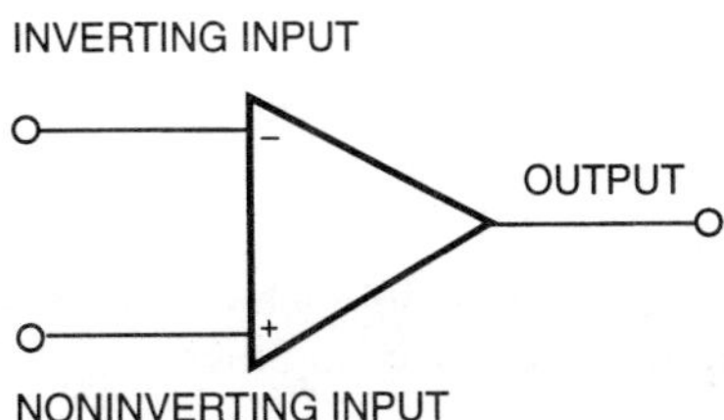

6. Define the term impedance.

It is convenient to represent the two elements of reactance, the magnitude and phase angle, in such a way that the result of combining several resistances and reactants can be determined easily. Thus, the series combination of a resistance, R, and an inductive reactance, X_L, is written as the complex impedance Z so that $Z = R + -1 X_L$.

7. What is the difference between the inverting and the noninverting input?

The operational amplifier consists of three gain units: a differential input unit, a high gain intermediate unit, and a high-power, low-impedance output unit. The differential input stage has two inputs, the inverting input $(-)$ and the noninverting input $(+)$. A positive output is given when a small positive voltage is applied to the noninverting output $(+)$. If a positive voltage is applied to the inverting input $(-)$, a negative output results.

8. What is "ground" for the operational amplifier?

The operational amplifier has no ground connection. The circuit works with split positive and negative supplies, e.g., $+20$ and -20. A grounded signal reference lies between the positive and negative voltages.

9. How do we operate in a differential mode with the operational amplifier?

To be in a differential mode, both inverting and noninverting inputs are used.

10. How do voltage amplifiers work? What are the essential principles that cause an input to be increased at the output? (Use the vacuum tube as an example.)

The advent of electronics is reckoned from the discovery that the current in a vacuum diode can be controlled by including a third electrode, the grid, in the space between the anode and the cathode. Very little power is consumed by the grid in controlling the anode current, so that the triode is an effective amplifier.

11. What is a grid?

A grid is a certain type of electrode.

12. What is a plate?

A plate is a flat- or grid-formed anode of an electron tube at which electrons collect.

Answers to Experiment 3 Questions

1. From a comparison of the zeroth, first, and second derivative excitation spectra of reserpine and rescinnamine, what derivative and wavelength are suitable for an analysis of an equal mixture of the two drugs?

To assay for reserpine, obtain the second derivative excitation intensity of the mixed standards at about 276 nm using the following instrument parameters: monitored λ_{ex}, 276 nm; maximum λ_{em}, 353 nm; scan speed, 60 nm/min; em slit width, 5 nm; ex slit width, 5 nm; sample sensitivity, coarse, 1, fine, 1; recorder speed, 2 in./min; recorder range, 100 mV; second derivative excitation scan range, 250 to 300 nm. Obtain the net intensity by measuring the perpendicular distance from the second minimum to the maximum peak at about 276 nm. Prepare a standard curve by plotting the points on linear paper.

To assay for rescinnamine, obtain the first derivative emission intensity of the mixed standards at about 411 nm using the following parameters: monitored λ_{em}, 411 nm; maximum λ_{ex}, 312 nm; scan speed, 60 nm/min; em slit width, 5 nm; ex slit width, 5 nm; sample sensitivity, coarse, 3, fine, 6; recorder speed, 2 in./min; recorder range, 100 mV; first derivative emission scan range, 340 to 500 nm. Obtain the net intensity as described for reserpine. Prepare a standard curve by plotting the points on linear paper.

2. From inspecting the spectra, can derivative luminescence be used for qualitative identification? How?

Derivative luminescence might be used as an aid in the qualitative identification of a metabolite from the original compound because a significant difference may exist in their zeroth, first, and second derivative spectra. The spectra may be similar, but are not identical, and consequently can be distinguished from one another.

3. *If only a filter fluorometer is available, name the method used to pick the correct excitation and emission filter for the second derivative maximum emission peak of reserpine.*

The maximum emission of the second derivative of reserpine in methanol occurs with the parameters λ_{ex} = 299 and λ_{em} = 350 nm. These parameters indicate the need for an excitation filter that will block the wavelength below about 295 nm and above about 305 nm. They also necessitate an emission filter that blocks emission wavelengths below about 345 nm and above about 355 nm.

4. *Name at least two metabolites of reserpine.*

Metabolites of reserpine include syringic acid, methyl reserpate, and 3,4,5-trimethoxybenzoic acid.

5. *Name several analytical problems that could be solved by the derivative luminescence technique.*

The derivative luminescence technique appears to be applicable to a variety of analytical problems and samples such as cross-contamination of drugs, medicated feeds, residues in animal tissues, and mixtures of mycotoxins or other mixtures that likely produce overlapping fluorescence bands.

6. *What is an oscilloscope?*

An oscilloscope is an instrument used for studying complex waveforms by displaying voltage waveforms visually.

7. *What are the uses for an oscilloscope?*

An oscilloscope may be used, for example, for displaying voltage waveforms, measuring the phase angles between sine waves, and displaying very high frequency signals.

8. *Define the term* phase angle *or* phase difference.

An a-c signal is a fraction of a cycle away from a reference point on the waveform at any instant. This is known as *phase*. The difference in phase of two signals is measured in degrees or in fractions of a complete angular rotation and is called the phase angle.

9. List two different methods of determining phase angles by instruments.

When a second signal is connected to the horizontal amplifier in place of the sweep generator, the oscilloscope becomes an x-y plotter displaying the functional relationship between the horizontal and vertical input signals. This mode of operation is used for measuring the phase angle of the two input signals. If two signals of the same frequency are connected to the horizontal and vertical inputs, it is possible to determine the phase relationship of the two signals.

10. In what way can a cathode ray tube (CRT) be useful in a television set?

The variation of grid to cathode voltage on the CRT changes the intensity of the trace on the screen. Intensity modulation of a CRT can be a useful variable in screen instrumentation and in television.

Answers to Experiment 4 Questions

1. Why don't ionic compounds dissolve to an appreciable extent in such solvents as benzene or carbon tetrachloride?

When an ionic salt is dissolved in water, the process of solution involves separation of the cations and anions of the salt with accompanying position of molecules of solvent about the ions. Such positioning of solvent molecules about the ions of the solute is possible only when the solvent is extremely polar and the dipoles of the solvent are drawn to and held by the ions of the solute. Benzene and carbon tetrachloride are nonpolar solvents.

2. Taking into account the results of this experiment, what was the effect on the wavelength of excitation and emission of reserpine as the dielectric constant of the solvent increased?

As the dielectric constant increased, the wavelengths of excitation and emission decreased.

3. **Based on the experiment, what was the effect on the wavelength of excitation and emission of rescinnamine as the dielectric constant of the solvent increased?**

As the dielectric constant of the solvent increased, the wavelength of excitation decreased and the wavelength of emission increased.

4. **Examine the structure of reserpine and rescinnamine and discuss possible reasons why the greatest relative fluorescent intensity for reserpine occurred in 1-butanol, while the greatest relative intensity for rescinnamine occurred in methanol.**

Reserpine has several functional groups on its rings, but the most common is the methoxide group, which generally has very little effect upon wavelength or intensity of emission unless steric factors are involved.

Rescinnamine not only has the methoxide group ($-OCH_3$) attached to the ring, but it also differs from reserpine in that attachment of the 1,2,3-methoxybenzene to the multiple ring system is by a $-OC-CH=CH-$ linkage in rescinnamine, while the linkage in reserpine is $-O-C-$. The effect of more than one substituent on fluorescence is thus a resultant effect upon the mobility or freedom of the π electrons.

Answers to Experiment 5 Questions

1. **Why does faster scanning speed result in a more intense fluorescent peak for a given slope?**

Increasing the scanning speed increases the rate of change of absorbance with wavelength (first derivative). The fastest moving rates of change are found near the peak maximum, therefore, this part of the absorption band exhibits the most vivid changes in the derivative spectrum.

2. **How would increasing the scanning speed aid the analysis of a sample in which only a small amount of sample was available?**

The scanning speed influences the intensity of the second derivative spectrum. A more rapid scanning speed produces a more rapidly changing absorption signal to the derivative circuit as well as a greater result. These conditions may permit analysis of a small amount of sample.

3. In the zeroth excitation and emission spectra, which peaks should give the greatest second derivative peaks?

Second derivative spectra turn broad peaks found in the absorption spectra into fine second derivative peaks.

4. If analysis was necessary for two fluorescent compounds having maximum peaks at the same wavelength, but different slopes in obtaining the peaks, how could the difference in slope along with increasing scan speed be used to find the compound with the greatest slope?

Second derivative spectra turn broad peaks found in the absorption spectra into fine second derivative peaks. The second derivative spectra often are able to resolve overlapping absorption bands. The selection of scanning speed also influences the intensity of the second derivative spectrum. A standard solution of the two compounds could be devised, and experiments in which different scanning speeds are used could be the means of determining the parameters within which the compound with the greatest slope could be determined.

Answers to Experiment 6 Questions

1. For which type of turbid solutions would derivative spectrofluorometry be used?

Turbid solutions that may be analyzed by derivative spectrometry include phenol in wastewater, quinine in fruit extracts, caffeine in dark-colored cola drinks, and the determination of protein in milk.

2. Given that vol $= 4/3\,\pi(r)^3$, find the volume of a particle that has a radius of 0.1 μm.

$$V = 4/3(3.14)(0.1\mu M)^3 = 0.004\ \mu M^3$$

3. Given that particles present in a solution have a volume of 6×10^{15} cm³ and density of 2 g/cm³, calculate the number of particles found in 5 g of the particles in a solution.

Mass of one particle = density × volume = 2 g/cm³ $(6 \times 10^{-15}$ cm³) = 12×10^{-15} g

No. of particles in 5 g = mass of all particles × 1 particle/12×10^{-15} g

$$= 0.42 \times 10^{15} \text{ particles}$$

4. What is the cross-sectional blocking area to a beam of monochromatic light passing through a solution containing 50×10^{13} particles in which each particle has a radius of 10^{-5} cm?

Cross-sectional blocking area = Area of 1 particle × no. of particles

$$= (\pi r^2) \times \text{no. of particles}$$

$$= 3.14(10^{-5} \text{ cm})^2/\text{particle} \times 50 \times 10^{13} \text{ particles}$$

$$= 157.1 \times 10 \text{ cm}$$

5. It is important to measure the concentration of a dissolved fluorescent solute, but present in the solution is a cloudy precipitate of some other solute. Filtration and centrifugation fail to remove this precipitate. Devise a means of determining the concentration of the dissolved fluorescent solute in the presence of the precipitate.

Assume that the cloudy precipitate can be identified. Prepare solutions using the same solvents in which the sample is dissolved to make up 0, 0.25, 0.5, and 1.0 g/l of the cloudy precipitate. Compare the cloudiness of the sample solution and determine the approximate concentration of the cloudy precipitate. If necessary, make up cloudy precipitate solutions at concentrations higher than 1.0 g/l. For example, let us assume that the cloudy precipitate has a concentration of about 0.5 g/l. Make up standard solutions of the fluorescent solute to contain 0, 2, 4, 6, and 8 ppm, with each solution containing 0.5 g/l of the cloudy precipitate. Obtain the zeroth, first, and second derivative spectra of each of the standards and the sample. If necessary, dilute the sample so it will fall within the standard curve. Use the second derivative spectra to quantitatively determine the concentration of the fluorescent solute.

6. Define Rayleigh and Raman scattering.

Within the size range of particles that cause scattering, for particles whose longest dimension is no more than 5 to 10% of the incident radiation, this scattering is known as Rayleigh scattering. Rayleigh scattering produces an intensity pattern of scattered radiation that is relatively symmetrical in all directions around the particles. The oscillating electric and magnetic forces comprising the incident light induce a dipole moment in molecules; they in turn radiate light of the same frequency (elastic scattering) as the incident or exciting radiation, but in all directions. Rayleigh scattering always accompanies Raman scattering. Some of the energy of the incident light in Raman scattering may be used in exciting a molecule to a higher vibrational or rotational energy level, and the radiation emitted (scattered) by the molecule is of correspondingly lower energy and frequency.

Answers to Experiment 7 Questions

1. Propose a derivative luminescent method for the analysis of a 1:1 phenanthrene-anthracene mixture in cyclohexane.

From the literature find the λ_{ex} (UV absorption maxima) for both phenanthrene and anthracene in methanol. Obtain a derivative spectrofluorometer and the zeroth, first and second derivative spectra of both anthracene and phenanthrene. Examine the excitation and derivative emission maxima and select wavelengths that will permit anthracene and phenanthrene peaks that do not interfere with each other. Make up 1:1 solutions of anthracene and phenanthrene and run excitation and emission scans. Prepare standard curves with 1:1 anthracene-phenanthrene. Now prepare the unknown, obtain scans, and determine the concentration of the analytes in the unknown.

2. Name a natural product that contains phenanthrene.

Phenanthrene occurs in coal tar.

3. *Name some of the physical and chemical properties of phenanthrene.*

Its density at 25°C is 1.179; melting point is 100°C; boiling point is 340°C; insoluble in water; soluble in organic solvents, forms molecular compounds with picric acid, picryl chloride, dinitrobenzene, and similar nitro compounds; and a solution of the analyte exhibits a blue fluorescence.

4. *Phenanthrene solutions exhibit a blue fluorescence. Using the information found in this experiment, to which wavelength does this correspond?*

The blue fluorescence probably corresponds to an emission wavelength of about 350 nm.

5. *What is an important use of anthracene?*

Anthracene is an important source of dyestuff.

Answers to Experiment 8 Questions

1. *Besides derivative spectrometry, name three other methods by which organic compounds are identified.*

(1) If the organic compound is a solid, it can be separated and purified, and its melting point obtained; (2) if the organic compound is a liquid, density, specific gravity, and boiling point help in identification; (3) suitably prepared, IR spectrometry of the sample, when compared to a known standard, aids identification. Other methods include GLC, thin-layer chromatography (TLC), and UV absorbance spectra for applicable compounds.

2. *Why do so many compounds have similar ultraviolet spectra?*

Because many compounds have a similar structure, such as the benzene ring, the prominent absorption band will be similar, although not exactly the same.

3. *When scanning speeds change, different derivative spectra are obtained. Why?*

The effect of scanning speeds on derivative spectra is based on the fact that derivatives are dependent on the rate of change of an absorbance spectra with respect to wavelength or time. Assume that a spectrometer sees an absorbance curve change from wavelength 1 to wavelength 2 in 0.1 s. This scanning speed will not give as sharp a derivative spectra as the condition in which the spectrometer scans from wavelength 1 to wavelength 2 in 0.01 s. The rate of change in the second case will be ten times faster and lead to a more pronounced derivative spectra and probably sharper peaks.

4. Name two or three other compounds that have similar UV spectra but that might possibly be identified using derivative spectroscopy.

Several compounds that have similar UV spectra but could be identified by derivative spectrometry include aspirin and salicylic acid, phenobarbital and amobarbital, and benz (α) anthracene in the presence of chrysene.

5. Name several uses of the compounds acenocoumarol, phenprocoumon, and warfarin.

All are anticoagulants in cases in which it is necessary to depress hepatic production of prothrombin. They are used to treat pulmonary embolism (to prevent further embolism), primary, acute, and postoperative thrombophlebitis, and traumatic injuries to blood vessels.

6. Suggest how derivative spectrometry may be used to quantitate the compounds discussed in this experiment.

Derivative spectrometry can be used to quantitate phenprocoumon,warfarin, and acenocoumarol in the following ways: (a) obtain a derivative spectrometer, (b) make up suitable separate standards of the analytes using 95% ethanol, (c) prepare samples in the same way as the standard was prepared and filter when necessary, (d) set up the derivative spectrometer and choose those parameters of scanning speed, wavelengths of excitation and emission, and other instrument parameters such as slit width, sensitivity settings, and recorder speed and recorder range that are able to obtain the best derivative spectra, (e) obtain a derivative spectra of a blank, samples, and standards using the same instrument parameters, (f) obtain peak height measurement of standard and sample at a chosen wavelength, and (g) calculate the quantity of ingredient in the sample using standard quantitative methods.

Answers to Experiment 9 Questions

1. Describe methods other than derivative spectrometry in which mixtures of sodium salicylate and acetaminophen can be analyzed.

No official method exists. However, a suggested method may be column chromatography, in which sodium salicylate is trapped on Celite 545 saturated with sodium bicarbonate while acetaminophen (in chloroform) is eluted off the column. The sodium salicylate would then be eluted into another volumetric flask using a weak acetic acid solution. Then, both compounds can be determined by UV absorption spectrometry using standards.

2. Could derivative fluorescence be used to analyze mixtures of acetaminophen and sodium salicylate? Why or why not?

Another possible method of analysis for sodium salicylate is derivative fluorescence analysis because the salicylates are strongly fluorescent, while acetaminophen is probably weakly fluorescent because it contains the hydroxy and $-NHCOCH_3$ groups.

3. Name several mixtures of compounds that you think could not be analyzed by derivative spectrometry and explain why.

Derivative spectrometry is used with compounds that absorb energy and give a UV or visible spectrum. Compounds that do not absorb energy to give a spectrum would not be applicable for analysis by derivative spectrometry. These would include many inorganic compounds such as iron oxide, sodium chloride, water, and calcium sulfate.

4. Using a Cary 118 spectrophotometric manual or some other manufacturer's derivative spectrometric manual, draw a schematic of a derivative spectrometer. Label all the parts.

For the answer, see the diagram on page 189.

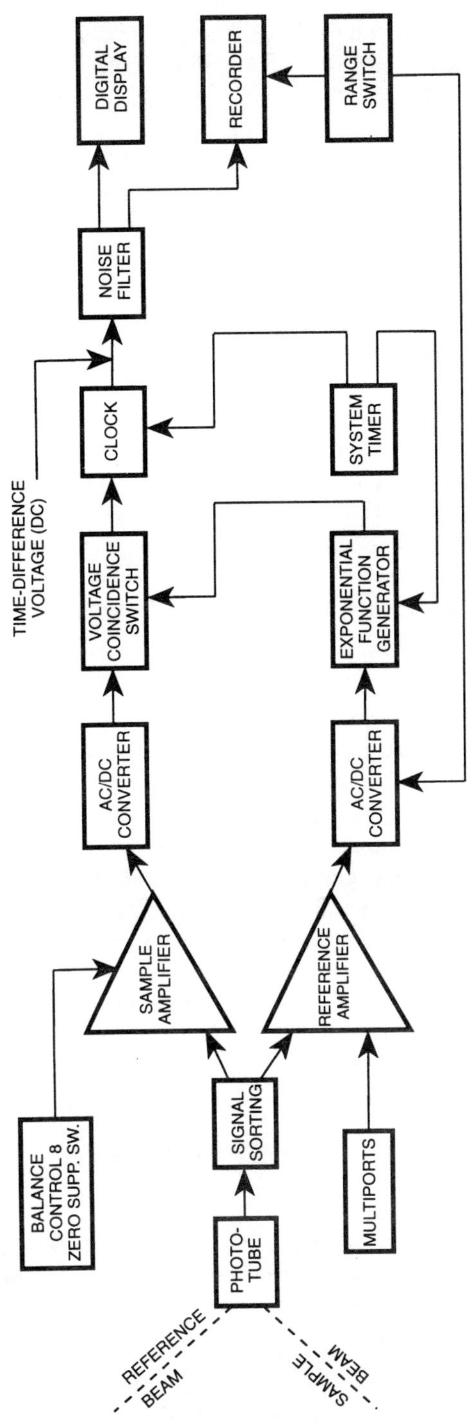

Answer to Question #4 of Experiment 9.

5. *Assume you have a solution containing 36 mg acetaminophen and 60 mg of sodium salicylate in 100 ml of 95% ethanol. Of this, 5.0 ml is taken and diluted to 50.0 ml with 95% ethanol. Of this solution, 5.0 ml is taken and diluted to 100.0 ml with 95% ethanol. What is the final concentration of acetaminophen and sodium salicylate in micrograms per milliliter?*

Acetaminophen would have a final concentration of 1.8 μg/ml and sodium salicylate would have a final concentration of 3 μg/ml.

6. *Tablets that are given to patients contain ingredients other than acetaminophen and sodium salicylate, such as diluents, colors, lubricants, disintegrants, starches, and lactose. Many of these ingredients are insoluble in ethanol. When tablets are analyzed by the derivative method described, what steps should probably be taken before the derivative spectrum of the sample is secured?*

In order to rid the tablet of excipients, most of which are water and 95% ethanol insoluble, weigh a portion of composite sample equivalent to approximately one average tablet weight, and place it in a 500-ml volumetric flask. Add 50 ml of ethanol and magnetically stir for 15 min on a hot plate at approximately 60°C. Carefully add 50 ml of boiling water and continue stirring for 15 min. Allow to cool and dilute to volume with water and mix well. Filter the solution through a suitable filter paper, discarding the first 20 ml of filtrate. Transfer a suitable amount to another volumetric flask and analyze for sodium salicylate and acetaminophen using derivative spectrometry. (Both the sodium salicylate and acetaminophen will be dissolved in the 1:1 95% ethanol-hot water solution.)

Answers to Experiment 10 Questions

1. *Name at least one other compound not having a pronounced UV maxima that could be analyzed using derivative spectroscopy.*

Because compounds with similar structures tend to give similar UV spectra and similar (but not identical) derivative spectra, a very good possibility exists that phenytoin and phenylethylhydantoin would also not have very distinct UV absorption spectra maximas and could be analyzed by derivative spectroscopy.

2. Why does the derivative of the phenytoin sodium UV spectra give distinct maxima while the UV absorption spectra do not?

The reason a compound such as phenytoin sodium can give pronounced derivative spectra even though its UV absorption spectra has no distinct maxima is due to the nature of derivative spectra. The derivative spectra is a function of the rates of change of an absorption spectra with respect to time or wavelength. Scanning speed also affects the rate.

3. The first derivative peak height of a sample of an impure phenytoin sodium was 25 mm at 258 nm while the corresponding peak height of a standard solution was 24 mm. If the standard concentration was 100 µg/ml and the sample consisted of 1.5 g of impure phenytoin sodium diluted to 100 ml in distilled water, calculate the concentration of phenytoin sodium in the sample and the percentage of phenytoin sodium in the sample.

Concentration of phenytoin sodium in the sample µg

$$= 25 \text{ mm}/24 \text{ mm} \times 100 \text{ µg/ml} \times 100 \text{ ml} = 10,400 \text{ µg}$$

% Phenytoin sodium in sample $= (10,400 \text{ µg})(1 \text{ mg}/1000 \text{ µg})$

$(1 \text{ g}/1000 \text{ mg})(1/0.15 \text{ g})(100) = 0.693\%$

4. A student mistakenly weighed 0.25 g of phenytoin sodium into a 10-ml volumetric flask and diluted to 10 ml with water. What was the concentration in micrograms per milliliter? How would the student take this solution and prepare a 100-µg/ml solution in water?

The student had a concentration of 25,000 µg/ml phenytoin sodium. To prepare a 100-µg/ml solution, place 1 ml of the 25,000-µg/ml solution into a 250-ml volumetric flask and dilute to 250 ml with water and mix well.

5. *Phenytoin and phenytoin sodium both give similar first derivative spectra. If you had only phenytoin standard and you wished to analyze for phenytoin sodium, how would you accomplish this?*

The following method should be used to analyze for phenytoin sodium using the phenytoin standard:

- Prepare a solution of phenytoin standard in 0.1 *N* NaOH at a concentration of about 125 µg/ml and mix well.

- Weigh out a sample equivalent to 1000 µg (10 mg) phenytoin sodium into a 100-ml volumetric flask and dilute to volume with 0.1 *N* NaOH. Mix well and filter if necessary.

- Obtain the first derivative spectra of each solution using the method described in the experiment.

- Calculate the concentration of phenytoin sodium in the sample in the following way:

$$\% \text{ Phenytoin sodium in sample} = \frac{\text{Peak height of sample}}{\text{Peak height of standard phenytoin}} \times \frac{\text{Molecular weight of phenytoin sodium}}{\text{Molecular weight of phenytoin}}$$

$$\times \text{ Conc. standard of phenytoin (µg/ml)} \times \frac{\text{Final volume of sample (ml)}}{\text{Weight of sample (g)}}$$

Answers to Experiment 11 Questions

1. *What are the advantages of being able to analyze for a drug without separation of interfering excipients?*

The advantages are that the analyst can avoid the cumbersome sample clean up and separation procedures and thus more rapidly complete the analysis.

2. *Name three other drug preparations for possible analysis by derivative spectroscopy without separation from excipients.*

Phenobarbital, 0.015 g in a 3.24-g suppository containing dipyrone (1 g); anisotropine methyl bromide (0.020 g) with other active ingredients and 2.204 g saturated fatty acid glycerides as excipients; a 150-ml solution of 1.50 g oxyfederine hydrochloride with 113.2 g excipients; and a solution containing estradiol benzoate (10 mg/ml) along with a suspension of lecithin (8 mg/ml) and chlorobutanol (3 mg/ml).

3. What methods other than derivative spectroscopy have been used to correct for background interference?

The practical limitation derived from variable and nonspecific spectral interference has been treated by graphic and mathematical methods. Some of the methods were very simple and others were very involved procedures in accordance with the shape of the irrelevant absorption spectrum. The Morton-Stubbs three-point correction method has been used extensively, but the method assumes a linear irrelevant absorption spectrum. Background interference also has been corrected by the application of absorbance ratios and orthogonal function coefficients which are sensitive to instrumental parameters and difficult to apply if the irrelevant background absorption is similar to the spectral band.

4. Using the USP method to analyze for edrophonium chloride, the net peak height for the sample was 75 mm, while the net peak height of the standard was 74 mm. Assuming a standard concentration of 55 µg/ml, calculate the milligrams per milliliter found in the edrophonium chloride injection if a 10-ml sample is taken for analysis.

mg/ml = (75 mm/74 mm) (55 µg/ml) (1 mg/1000 µg) (100 ml/5 ml) (50 ml/10 ml) edrophonium chloride.

5. Using the derivative analysis described in this experiment, calculate the milligrams per milliliter of edrophonium chloride in the injection if given the following data: net peak height sample, 50 mm; net peak height sample, 49 mm; concentration standard, 44 µg/ml; ml original sample taken for analysis, 10 ml; and aliquot taken from 250-ml sample solution, 10 ml.

mg/ml = (50 mm/49 mm) (44 µg/ml) (1 mg/1000 µg) (250 ml/10 ml) (50 ml/10 ml) edrophonium chloride = 5.61 mg/ml.

Answers to Experiment 12 Questions

1. Name one other steroid mixture that could be analyzed using derivative spectrometry.

6α-Methylprogesterone can be analyzed in the presence of 6β-methylprogesterone.

2. In your opinion, what causes the intense maximum at about 300 nm in the derivative spectra of the 6α-methyl testosterone compounds that is not found in the 6β-methyl testosterone?

The origin of the intense maximum at about 300 nm in the derivative spectra of the 6α-methyl series is not known at this time.

3. Name two other steroids (besides testosterone) found in humans and their uses.

These steroids are progesterone and androsterone. Progesterone appears to have the highest specificity of the sex hormones. Ovulation and pituitary function are dependent on the cyclic variations of estrogen and progesterone. Androsterone is a male sex hormone and contributes to the formation of male secondary characteristics such as a deep voice, facial hair, sexual drive, and well-developed penis.

4. If you had a mixture of the two steroids mentioned in Question 3, propose a possible derivative spectrophotometric analysis of one of them in the presence of the other.

A derivative spectrophotometric analysis of combined progesterone and androsterone might be as follows: let us assume that the second derivative spectra of a mixture of these compounds (5 mg/ml each in diethylene glycol dimethyl ether) showed a maximum peak at 300 nm for progesterone, while androsterone was at a minimum peak. Let us further assume that at 260 nm androsterone gave a maximum peak while progesterone was at a minimum. Then a sample mixture and standard mixtures of about the same concentrations could be weighed out into an appropriate volume of diethyl glycol dimethyl ether and the derivative spectra obtained. The quantity of progesterone and androsterone in the sample could then be calculated.

5. *If a mixture of 6α- and 6β-methyl testosterone gave a net derivative peak of 25 nm at 300 nm, while a corresponding standard mixture containing 5.2 μg/ml 6α-methyl testosterone and 4.5 mg/ml 6β-methyl testosterone gave a net peak height of 24 nm for α-methyl testosterone under the same circumstances, calculate the concentration of 6α-methyl testosterone in the sample. Assume similar volumes of the sample and the standard.*

Concentration of 6α-methyl testosterone in sample = 25 mm/24 mm × 5.2 μg/ml = 5.4 mg/ml.

Answers to Experiment 13 Questions

1. *How would you define a first derivative spectrum?*

This is defined as the limit of the difference quotient, $\Delta y/\Delta x$, of two simultaneous changes Δx and Δy in two independent variables, x and y, as the increment Δx approaches zero is called the derivative of the dependent variable y with respect to the independent variable x at the value of x in question.

2. *What is meant by the term "second derivative"?*

The second derivative is the derivative of the first derivative.

3. *Why do you think saccharin tastes sweet?*

4. *Why does sugar taste sweet?*

"Sweet" is defined in *Webster's Dictionary* as having a taste similar to that of sugar and having a generally agreeable taste, smell, and appearance. Saccharin tastes like sugar in most cases.

5. Is there any common chemical characteristic between saccharin, sugar, and aspartame (the ingredient found in NutraSweet®) that would lead one to believe that certain chemical structures are associated with sweet taste?

There is no apparent common chemical characteristics or common chemical structures with regard to aspartame, saccharin, or sugar.

6. Why has aspartame become more popular in usage than saccharin?

Aspartame appears to be more popular than saccharin (i.e., used in most colas, commercial drinks, etc.) because saccharin in high doses was found to cause cancer in laboratory animals, and the FDA, following the Delaney Amendment to the Food, Drug and Cosmetic Act, must ban any cancer-causing product. However, Congress blocked the FDA's action because diabetics needed a sugar substitute.

7. If a 5-μg/ml saccharin standard gives a first derivative peak of 25 mm when a 100-ml solution of a cola sample containing saccharin gives a first derivative peak of 28 nm at the same wavelength, what is the concentration of saccharin in the sample if the 100-ml solution of the sample contains 5 ml of the original cola?

Concentration of saccharin = 28 mm/25 mm $\times$ 5 μg/ml $\times$ 100 ml/5 ml = 112 μg/ml in cola sample.

Answers to Experiment 14 Questions

1. Name five other gases that can be analyzed by derivative spectroscopy.

These gases are nitric oxide, nitrogen dioxide, sulfur dioxide, ozone, and benzene vapor.

2. Propose a derivative method that can be used to analyze for one of the gases mentioned in Question 1.

An analysis method for sulfur dioxide which has large derivative peaks at about 302 nm would involve obtaining the second derivative spectra of several standard solutions and a sample solution between 325 and 280 nm, preparing a standard curve, and determining the amount of sulfur dioxide in the sample. The use of a gas cell and a method that introduces a representative portion of gas to the cell are needed.

3. Give a brief explanation of the particle and the wave nature of light.

Light appears to have a particle nature in that absorption of light occurs when these particles, called photons or quanta, collide with an atom, molecule, or ion and increase their potential energy. In terms of the wave nature of light and other radiation, the absorption of light arises mainly from the interaction of the electric vector with a electron. The interaction of the magnetic vector with an electron is small enough to be ignored. Absorption of light increases the velocity of an electron primarily by changing its direction, not its speed. Velocity is a vector quantity that is specified by direction as well as speed. The direction and thus the velocity of an electron is changed when its orbital angular momentum is changed. Therefore, the electric vector of absorbed light can change the angular momentum of an electron in an s orbital to that of a p orbital. Similarly, it changes the angular momentum of an electron in a p orbital to that of a d orbital, etc.

4. What is the relationship between the angstrom, the millimicron, and the nanometer?

10 Å = 1 millimicron = 1 nanometer.

5. Define the electronic ground state and the electronic excited states of atom, ion, or molecule.

The *electronic ground state* is that in which all of the electrons of the species are in their most stable orbitals. An *electronic excited state* is one in which at least one electron occupies a higher energy orbital than it does in the ground state. Thus, the electron has a higher potential energy than it did in the ground state. A substance may have a number of excited states; for an atom, these may be simply designated as A_1, A_2, A_3, etc. Similarly, the ground state of a atom may be designated as A_0.

Answers to Experiment 15 Questions

1. **The rare earths constitute as a group only 0.008% of the weight of the Earth's crust. How much rare earth in grams is found in 1 million lb of that crust?**

   1 million lb $\times$ 0.0008 $\times$ 454 g/lb = 36,320 g.

2. **Gadolinium has a density of 7.886 g/ml. If a sample of pure gadolinium has a volume of 1.64 cm³, what is its weight?**

   Weight (g) = 7.886 g/ml $\times$ 1.64 ml = 12.9.

3. **Some pure metal samples contained in beakers lost their labels. One sample was found to have a volume of 2 ml and weighed 14 g. The metal was thought to be gadolinium, europium, samarium, or promethium (densities of 7.886, 5.245, 7.536, and 7.016 g/cm³, respectively). What was the identity of the metal?**

   Density = 14 g/2 ml = 7 g/ml. This density is closest to that of promethium, which is 7.016 g/ml; however, this tentative identification would necessitate verification by other tests.

4. **Describe the atomic structure of gadolinium, including its outer electron configuration, and explain how gadolinium undergoes chemical reactions.**

   Gadolinium has an atomic number of 64 with a probable outer electronic configuration of $-4f^7\ 5d^1\ 6s^2$. The entire electronic structure of gadolinium is [Xe] $4f^7\ 5d^1\ 6s^2$. Gadolinium has a valence of 3. When the oxide of gadolinium is heated it reacts with ammonium chloride to form the chloride. It also forms sulfates, nitrates, hydroxides, and oxides under appropriate conditions.

5. **Gadolinium has the following properties: atomic number, 64; atomic weight, 157.25; probable outer electronic configuration, $4f^7\ 5d^1\ 6s^2$; density, 7.886 g/ml; atomic volume, 19.88 ml/mol; melting point, 1312°C;**

and boiling point, 300°C. Name a method that could be used to separate gadolinium from a mixture of samarium and europium.

Gadolinium, samarium, and europium have boiling points of 3000°, 1439°, and 1804°C, respectively. This great divergence in temperatures may be used to selectively boil and convert to a gas each of the elements in turn, with the gas being suctioned off into separate flasks and then cooled to obtain the pure metal.

6. *Gadolinium nitrate absorbs maximally at about 272 nm in broad peak. Draw a hypothetical absorption spectra for this compound between 350 and 250 nm.*

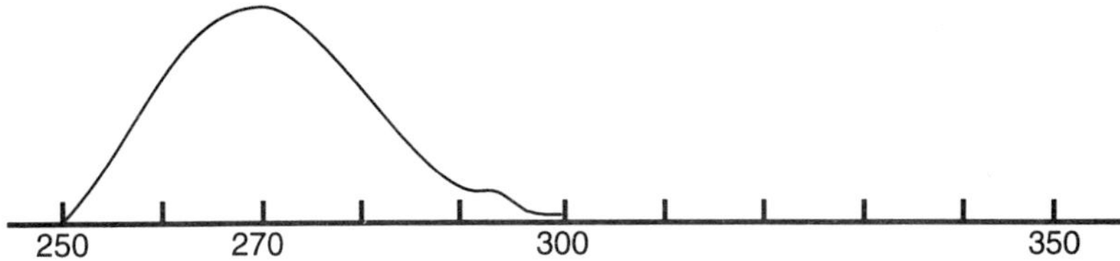

Answers to Experiment 16 Questions

1. *Phenacetin can be analyzed by a variety of methods. Describe the official method found in the USP.*

The old USP method for phenacetin involves the extraction of phenacetin in a weighed portion of well-ground and mixed tablets using petroleum ether followed by chloroform. The solvents are evaporated off and the phenacetin weighed.

2. *Phenacetin contains the CH_3 group. Review an organic chemistry book and describe the type of bonds found with this type of carbon.*

The carbon on $-CH_3$ contains four sp^3 hybrid orbitals. Of these, three hybrid orbitals overlap the $1s$ orbital of hydrogen and the other sp^3 hybrid orbital overlaps the hybrid orbital of an adjacent carbon atom, if we assume that the $-CH_3$ group is attached to a carbon chain.

3. *Assume that you had some old phenacetin tablets that had obviously broken down. The tablets declare 250 mg per tablet. Give the general steps that you would follow in order to analyze these tablets.*

Weigh out 20 tablets and obtain the average weight. Grind the tablets and pass through an appropriate sieve. Weigh out the equivalent of 100 mg of phenacetin into a volumetric flask and treat as described in this experiment.

4. *Phenacetin breaks down into p-phenetidin and acetic acid under the influence of strong acids or bases. Write the equation (using condensed formulas) for this reaction.*

$$CH_3CH_2O-\!\!\!\bigcirc\!\!\!-NHCCH_3 \ \xrightarrow[H^+ \text{ or } (OH)^-]{H_2O} \ CH_3COH \ + \ CH_3CH_2O-\!\!\!\bigcirc\!\!\!-NH_2$$

Answers to Experiment 17 Questions

1. *Tell how derivative spectrometry can be used to analyze a mixture of vitamins D and E.*

One must first find suitable solvents and then obtain UV absorption and first and second derivative spectra. From examining these spectra a set of operating procedures can be devised to analyze for these compounds.

2. *Assume you have a solution of a vitamin that you suspect is pyridoxine hydrochloride. Tell how you could use derivative spectrometry to qualitatively identify this compound.*

To qualitatively identify pyridoxine hydrochloride, prepare a standard solution of pyridoxine HCl in phosphate buffer solution, pH 7, at 10 mg/ml and compare the second derivative spectra of the unknown similarly prepared to that of the standard.

3. Why should our daily diet be drawn from a variety of foods?

No single food contains all nutrients. People place their health in danger when they go on diets limited to one particular food.

4. On a strict vegetarian diet, what one vitamin is hardest to obtain?

Most patients with B_{12} deficiency are vegans.

5. Derivative spectrometry is based on absorption of energy. Explain the absorption process in terms of the wave nature of light.

In terms of the wave nature of light and other radiation, the absorption of light arises from the interaction of the electric vector with an electron. The interaction of the magnetic vector with an electron is small enough to be ignored. Absorption of light increases the velocity of an electron primarily by changing its direction, not its speed. The direction, and thus the velocity, of an electron is changed when its orbital angular momentum is changed. The electric vector of absorbed light can change the angular momentum of an electron in an s orbital to that of a p orbital or an electron in a p orbital to that of a d orbital.

INDEX

A

Abamectin, 114, 117–118
Absorption
  absorbance and, 8
  bands, in spectra, 30, 47, 67–70
  Beer's law and, 8
  electron energy levels and, 197, 201
  infrared, 126–129
Absorption cell, 77
Acenocoumarol, 48, 50
Acephate, 99
Acetaminophen, 51–56
  decomposition of, 84
  structure of, 55
Acetone, 26
Acetophenetidin, see Phenacetin
Acetylcholine, 66
Acidity, solubility and, 22, 26
Addison's disease, 70
Air, analysis of, 41, 146–150
Alcohols, solubility of, 25–26
Aldehydes, solubility of, 26
Aldrin, 99
Alkaloids, quantitation of, 40
Aluminum silicate, 34–35
American Instrument Co. Inc., 155
Aminco SPF-125 spectrofluorometer, 155
p-Aminophenol, 84
Ammeter, recording, 127
Ammonia, 76, 78–79
Ammonium sulfate, 77
Amplifiers, 177
Amplifiers, operational, see Operational
  amplifiers
Analgesic drugs, 56, 87
Analog Devices, 10
  Model 174 operational amplifier
    manifold, 7
Anderson, R. C., 144, 174
Angstrom, 197
Anhydrides, solubility of, 26
Anorexia, 96
Anthracene, 10, 11, 42, 186
Anticoagulant drugs, 49, 50
Anticurare agent, 65–66
Antihypertensive drugs, 16
Antipyretic drugs, 56, 87

Applied Biosystems, 159
Aramu, I., 130, 174
Ascarelli, G., 134, 174
Aspartame, 196
Atmospheric turbulence, 141–142
Atomic absorption spectrometry (AAS), 80
Atrazine, 99
Avermectin Bla and Blb, 114–118
Aviv Associates, 159
Azodrin, 99

B

Beam splitters, reflectance spectra and,
  150–152
Beckman Instruments, 160
Beer-Lambert law, 33
Beer's law, 8
Begutter, H., 118, 174
Benzene, in alcohol, quantitation of, 40
Benzoic acid, fluorescence of, 19
Beriberi, 94
Beverages, analysis of, 72–75
  turbidity and, 33, 40
Biotin, 96
Blocking area, cross-sectional, 184
Bonfiglioli, G., 130, 132, 174
Bordeaux mixture, 100
Breadboards, 3–4, 175
Brovetto, P., 130, 132, 174
Buck Scientific, Inc., 160
Busca, G., 132, 174

C

Capacitors, 175
Captan, 103–106
Carbamate insecticides, 99
Carbaryl, 99
Carbon dioxide, quantitation of, 40
Carbonyl compounds, fluorescence of,
  19–20
Cary Model 118 spectrophotometer, 47
Cary Products, 47
Castration, 70
Chemie Linz, 120
Chlorinated insecticides, 99
Chloroxuron, 107